About Island Press

Since 1984, the nonprofit organization Island Press has been stimulating, shaping, and communicating ideas that are essential for solving environmental problems worldwide. With more than 1,000 titles in print and some 30 new releases each year, we are the nation's leading publisher on environmental issues. We identify innovative thinkers and emerging trends in the environmental field. We work with world-renowned experts and authors to develop cross-disciplinary solutions to environmental challenges.

Island Press designs and executes educational campaigns in conjunction with our authors to communicate their critical messages in print, in person, and online using the latest technologies, innovative programs, and the media. Our goal is to reach targeted audiences—scientists, policymakers, environmental advocates, urban planners, the media, and concerned citizens—with information that can be used to create the framework for long-term ecological health and human well-being.

Island Press gratefully acknowledges major support from The Bobolink Foundation, Caldera Foundation, The Curtis and Edith Munson Foundation, The Forrest C. and Frances H. Lattner Foundation, The JPB Foundation, The Kresge Foundation, The Summit Charitable Foundation, Inc., and many other generous organizations and individuals.

Generous support for the publication of this book was provided by Margot and John Ernst.

The opinions expressed in this book are those of the author(s) and do not necessarily reflect the views of our supporters.

Urban Raptors

Urban Raptors

Ecology and Conservation
of Birds of Prey in Cities

Edited by Clint W. Boal and Cheryl R. Dykstra

ISLANDPRESS

Washington | Covelo | London

Library of Congress Control Number: 2017958112

All Island Press books are printed on environmentally responsible materials.

Manufactured in the United States of America.
10 9 8 7 6 5 4 3 2 1

Keywords: barred owl, behavioral ecology, burrowing owl, community ecology, conservation and management, Cooper's hawk, falcons, habitat management, Harris's hawk, hawks and eagles, human-wildlife conflict, incorporating modern technology, Mississippi kites, nest-site selection, ornithology, owls, peregrine falcon, population ecology, powerful owl, raptor mortality, raptor rehabilitation, red-shouldered hawk, spatial modeling, suburban wildlife, urban greenspace planning, urbanization, wildlife habitats, wildlife management

Contents

Preface   xi

PART I   Raptors in Urban Ecosystems   1

CHAPTER 1   Urban Birds of Prey: A Lengthy History of
Human-Raptor Cohabitation   3
    Keith L. Bildstein and Jean-François Therrien

CHAPTER 2   City Lifestyles: Behavioral Ecology of Urban Raptors   18
    Cheryl R. Dykstra

CHAPTER 3   Urban Raptor Communities: Why Some Raptors
and Not Others Occupy Urban Environments   36
    Clint W. Boal

CHAPTER 4   Demography of Raptor Populations in Urban
Environments   51
    R. William Mannan and Robert J. Steidl

CHAPTER 5   Urbanization and Raptors: Trends and Research Approaches   64
    Raylene Cooke, Fiona Hogan, Bronwyn Isaac, Marian Weaving,
    and John G. White

PART II   Urban Raptors   77

CHAPTER 6   Mississippi Kites: Elegance Aloft   79
Ben R. Skipper

CHAPTER 7   Cooper's Hawks: The Bold Backyard Hunters   93
Robert N. Rosenfield, R. William Mannan, and Brian A. Millsap

CHAPTER 8   Red-Shouldered Hawks: Adaptable Denizens of the Suburbs   110
Cheryl R. Dykstra, Peter H. Bloom, and Michael D. McCrary

CHAPTER 9   Harris's Hawks: All in the Family   126
Clint W. Boal and James F. Dwyer

CHAPTER 10   Barred Owls: A Nocturnal Generalist Thrives in Wooded,
Suburban Habitats   138
Richard O. Bierregaard

CHAPTER 11   Powerful Owls: Possum Assassins Move into Town   152
*Raylene Cooke, Fiona Hogan, Bronwyn Isaac, Marian Weaving,
and John G. White*

CHAPTER 12   Burrowing Owls: Happy Urbanite or Disgruntled Tenant?   166
Courtney J. Conway

CHAPTER 13   Peregrine Falcons: The Neighbors Upstairs   180
*Joel E. Pagel, Clifford M. Anderson, Douglas A. Bell, Edward Deal,
Lloyd Kiff, F. Arthur McMorris, Patrick T. Redig, and Robert Sallinger*

PART III   Conservation and Management   197

CHAPTER 14   Raptor Mortality in Urban Landscapes   199
James F. Dwyer, Sofi Hindmarch, and Gail E. Kratz

CHAPTER 15   Human-Raptor Conflicts in Urban Settings   214
Brian E. Washburn

CHAPTER 16 Raptors as Victims and Ambassadors: Raptor Rehabilitation, Education, and Outreach 229
Lori R. Arent, Michelle Willette, and Gail Buhl

CHAPTER 17 Urban Raptor Case Studies: Lessons from Texas 246
John M. Davis

CHAPTER 18 Management and Conservation of Urban Raptors 258
David M. Bird, Robert N. Rosenfield, Greg Septon, Marcel A. Gahbauer, John H. Barclay, and Jeffrey L. Lincer

CHAPTER 19 Perspectives and Future Directions 273
Stephen DeStefano and Clint W. Boal

CONTRIBUTORS 287

INDEX 293

Preface

Raptors, or birds of prey, are members of three distinct groups of birds: the hawks and eagles, the falcons, and the owls. Though not closely related in an evolutionary sense, these three groups possess similar features that set them apart from all other birds: intense, forward-looking eyes; a sharply hooked bill for tearing bites of food; and powerful, sharp talons to grab and subdue their prey. These features also make them immediately recognizable to most people. Even if casual observers don't know a peregrine falcon from a red-tailed hawk, they almost always feel a visceral thrill of knowing beyond doubt that they are looking at a "raptor," and know that, for some indescribable reason, it is somehow special.

This sentiment is not limited to the modern observer but is somehow ingrained in the human species, for both good and ill. Since the dawn of our first civilizations, humans have had a long and fascinating relationship with raptors. Initially, raptors served as symbolic representations of gods and divine power. The ancient Irish told the myth of the Hawk of Achill, the "grey hawk of time," one of the oldest and wisest of animals. Some cultures considered owls wise, such as the symbol of Athena/Minerva, the Greek/Roman goddess of wisdom, whereas others thought them to be omens of death or evil. In North America, the giant eagle or "thunderbird" was an important deity of many Native American tribes. Perhaps the most well-known is the falcon-headed Horus, the ancient Egyptian god of the sky. Raptors have also symbolized national might; the soldier honored with carrying the eagle standard for a Roman legion held the prestigious title of *Aquilifer*. Today, raptors serve as the national symbol for no fewer than 26 countries and as mascots for countless schools, colleges, and athletic teams.

Aside from the spiritual and symbolic connections and unlike almost any other wild animal, raptors have served humankind in a literal sense. Over 2,000 years ago, people learned to capture and train raptors as hunting partners to acquire food, a custom still practiced today through the sport of falconry. But this relationship between humans and raptors has been inconsistent. Once firearms were developed, raptors were perceived as pests and varmints, threats to livestock or competition for game, and the resulting persecution was merciless. Raptors became known as creatures of the remote and wild places, intolerant of human presence and activities. Today, we are learning that this reticence may have been due less to behavioral intolerance by the raptors than to intense persecution by humans. Somewhat surprisingly, in recent decades, some raptor species have moved into, occupied, and in some cases developed substantial populations in urban areas around the world.

There are several possible explanations for this unexpected coexistence, which we will explore in this volume. Raptors are avian carnivores, so within these urban settings they must find, capture, and kill their prey. But it appears that many are able to do so with minimal conflict with humans. Whereas mammalian carnivores, such as mountain lions or leopards, present a physical threat to humans or their pets, any risk from raptors is minimal. What is more, many mammalian carnivores are nocturnal and thus rarely seen or appreciated by humans. In contrast, many of the raptors found in urban settings are diurnal and can not only be seen by residents but also are actively sought out by birding enthusiasts as exciting examples of "watchable wildlife" in urban backyards and parks. For example, each spring thousands of people watch the widely known red-tailed hawks that have nested adjacent to Central Park in New York City since 1991; multiple books and a film documentary have been produced about these hawks. The streaming web camera focused on a nesting pair of bald eagles in Washington, DC, receives thousands of questions during live Q&A sessions. There is even a public Facebook page dedicated solely to urban raptors. The popularity of these urban-dwelling raptors is substantial.

The genesis of this book stemmed from conversations with our colleagues, often during our annual conferences of the Raptor Research Foundation, where we discussed our ongoing research and attended presentations by other researchers and students studying urban raptors. We realized that an incredible volume of research had been conducted on urban birds of prey in the last 20 years but had not yet been assembled into a single source of information. Within the scientific discipline of ornithology, researchers now know far more about birds of prey and their role as members of biological communities in the novel setting of urban ecosystems. This provides both opportunities and challenges for conservation, environmental education, and management. What is lacking, however, is a compilation of the existing knowledge on this topic. This volume is our attempt to fill that void by providing an overview of urban ecosystems in the context of raptor ecology and conservation. Our goal is to provide a valuable source of information for researchers, urban green space planners, wildlife management agencies, bird-watching enthusiasts, and interested citizens.

This volume is divided into three parts: Raptors in Urban Ecosystems (Part I), Urban Raptors (Part II), and Conservation and Management (Part III). Part I focuses on urban settings and how birds of prey function within them. We start this section with Keith Bildstein and Jean-François Therrien's overview of the history of human interactions with raptors and how changes in human behavior have contributed to growing populations of urban raptors. Cheryl Dykstra follows with an exploration of behavioral aspects of birds of prey that have allowed some species to capitalize on different components of urban settings. Clint Boal discusses how food habits and competition may influence the composition of urban raptor communities. Bill Mannan and Bob Steidl explore how urban raptors may experience greater or lesser reproductive success and population growth compared to those in non-urban areas. Finally, Raylene Cooke and her colleagues explain how attributes of urban settings can allow those settings to function as unique ecosystems that may be attractive to raptors and how modern research tools help us study and conserve urban raptors. All together, these chapters provide an understanding of the urban ecosystems, human actions, and raptor behaviors that facilitate the presence of some species, and not others, in cities. This is foundational to understanding the unique aspects of individual species and conservation and management issues in the subsequent sections.

It is beyond the scope of this volume to detail every raptor species that may be found in urban settings. Therefore, in Part II we provide accounts for a sample of species that are found occupying urban areas and are representative of unique aspects of urban raptors. We start with Ben Skipper's account of the Mississippi kite, a primarily insectivorous raptor that has actually expanded its range in concert with urban development. Bob Rosenfield, Bill Mannan, and Brian Millsap discuss the Cooper's hawk, a bird-hunting species that occupies many cities across

North America. Cheryl Dykstra, Pete Bloom, and Michael McCrary then describe the urban ecology of the red-shouldered hawk, a woodland hawk that is quite generalist in its food habits. Clint Boal and James Dwyer discuss the Harris's hawk, a behaviorally unique raptor that not only breeds but also hunts in family groups. Rob Bierregaard introduces the barred owl, a nocturnal generalist that makes its home in mature, wooded neighborhoods. From Australia, Raylene Cooke and her coauthors acquaint the reader with the powerful owl, a large nocturnal species that is dependent on similarly large tree cavities for nesting. The urban ecology of the burrowing owl, another cavity nester, but this one dependent on ground burrows, is described by Courtney Conway. Joel Pagel and his coauthors close out the species section with an overview of what is possibly the most well-known urban raptor and conservation success story, the peregrine falcon.

Equipped with an introduction to urban ecosystems and some familiarity with raptors and their behaviors, we advance to Part III and delve into conservation and management issues. James Dwyer, Sofi Hindmarch, and Gail Kratz begin this section by discussing how urban landscapes present an array of risks that urban raptors are unlikely to encounter in non-urban areas to any great extent. Brian Washburn expands on this by explaining the situations in which conflict can occur between humans and raptors in urban settings and how management efforts can be made. When bad things happen to good city raptors, however, humans often intervene: Lori Arent, Michelle Willette, and Gail Buhl discuss the increasing role of wildlife rehabilitators and how raptors also serve as wildlife "ambassadors" in environmental education programs. John Davis follows with case studies of real-world situations that wildlife managers have resolved in urban settings. David Bird and his coauthors provide an overview of the conservation and management of urban raptors from the researcher's perspective, and Stephen DeStefano and Clint Boal close this volume with a synthesis of the material provided and suggestions for future research and conservation efforts.

More than ever, we are becoming aware of how small our planet is. With over seven billion humans currently occupying it, there is literally no place on earth that we are not having direct and indirect influences on wildlife. Indeed, as wild places are rapidly lost to anthropogenic development, it can legitimately be argued that urban ecosystems are the only type of ecosystem that is increasing on the planet. At first glance, this does not bode well for wildlife, and for many species, this may be true. But there is another side, a glimmer of hope for maintaining wildlife in our ever-urbanizing world. Raptors are top-trophic-level predators that have unique nesting and habitat requirements and require a variety of other animals to hunt for food. The fact that some raptors occupy urban areas is evidence that these areas can function as ecosystems, providing habitat and food not only for raptors, but also for the animals upon which they prey. It is our hope that this volume may play a role in facilitating a greater understanding and appreciation of birds of prey and foster an increased willingness to accommodate them as important members, not intruders, of our urban ecosystems.

Acknowledgments

Many of the chapters in this volume were presented at the 50th anniversary meeting of the Raptor Research Foundation in October 2016 at Cape May, New Jersey. We thank the organizers and volunteers of that conference for the opportunity to bring together colleagues from all around the world for a special symposium dedicated to the ecology of urban and suburban raptors. Specifically, we thank Lillian Armstrong and David La Puma, cochairs of the local committee; New Jersey Audubon, Cape May Bird Observatory, and Cape May Raptor Banding Project, who hosted the meeting; and the numerous sponsors of the conference. We appreciate the work of scientific program chair Beth Wommack, who supported the symposium and managed abstract submission, as well as the team of volunteers who tirelessly addressed the inevitable technical difficulties at the meeting.

We are grateful to the referees who reviewed earlier drafts of the chapters: David Andersen, John Barclay, Brent Bibles, Karen Cleveland, Jennifer Coulson, Steve DeStefano, Michelle Durflinger, Allen Fish, Mark Fuller, Bill Mannan, Dave McRuer, Brian Millsap, Jeff Sipple, Lynne Trulio, and Katheryn Watson. We thank the authors for their commitment to studying urban raptors, for spending their days in urban and suburban environments that lack the glamour and beauty of remote, natural field sites, and for their enthusiasm for participating in the symposium and this volume. The chapters in this volume are illustrated by the work of several photographers; each is identified with their photographs, and we appreciate their gracious contributions that bring visual life to this volume. We are especially appreciative of Courtney Lix and Elizabeth Farry for their interest and enthusiasm for this project and for shepherding us through the publication process with Island Press.

Finally, we thank our families for their support and patience during the production of this book and the numerous field seasons that preceded it.

Clint Boal and Cheryl Dykstra

PART I

Raptors in Urban Ecosystems

CHAPTER 1

Urban Birds of Prey: A Lengthy History of Human-Raptor Cohabitation

Keith L. Bildstein and Jean-François Therrien

POPULATIONS OF "URBAN" RAPTORS ARE increasing globally. Trained falcons are now being flown in city golf courses to scare off geese in hopes of reducing accumulated droppings along the fairways. In both the Old World and New, tens of thousands of vultures rummage through urban garbage dumps in search of humans' leftovers. In Spain, lesser kestrels (*Falco naumanni*) raise their young in the center of cities and towns, where they are attracted to and feed on swarms of insects flying above night-lit cathedrals and other historic buildings. Peregrine falcons (*Falco peregrinus*) routinely hunt for birds attracted to the brightly lit Empire State Building in downtown New York City, and red-tailed hawks (*Buteo jamaicensis*) nest in and around Central Park, feeding on pigeons, rats, and squirrels. Many other species serve as additional examples of a growing number of "urban" birds of prey, whose populations are increasing as human attitudes shift from a "shoot-on-sight" mentality to indifference and tolerance. But before exploring this topic further, first we will offer a bit of linguistics to explore the nuances of the phrase "urban raptor."

The word *urban* is believed to be derived from the Latin word *urbs*, which refers to a "walled city" or, specifically, to ancient Rome. Today it is used to indicate areas with high-density human settlements and is defined in the fifth edition of the *Oxford English Dictionary* as being "of, pertaining to, or consti-tuting a city or town."[1] The word first came into use in the English language in the early 17th century, thousands of years after human cities themselves first appeared.

Although raptors, more than most birds, have been heavily persecuted by humans, there is evidence that "urban raptors" began to appear simultaneously with human-created urban landscapes. Indeed, relationships between raptors and humans—some commensal, some mutually beneficial, and others still par-asitic or predatory—probably predate modern humanity itself.[2,3] That said, most studies of urban birds,[4] including those of raptors,[5] have been conducted in the past 35 years, and as such, the serious study of urban raptors remains in its infancy, with some researchers suggesting that the phenomenon of "urban rap-tors" is relatively recent.

Nevertheless, there has been a lengthy buildup to the phenomenon of city birds of prey, highlighted by many kinds of symbiotic relationships between humans and raptors that predate and, in many ways, foreshadow this ongoing phenomenon. Here, we cast this relationship in the light of two well-established and closely related ecological principles: habitat selection and expanded niche breadth coupled with population growth. Specifically, habitat selection results in raptors settling in landscapes that provide them with both safe nesting sites and adequate and accessible feeding sites,[6] or in less technical terms, a safe "bedroom" and a well-stocked "pantry" or "kitchen" (an ecological connection that then US Secretary of the Interior Cecil Andrus made while proposing the expansion of the Snake River Birds of Prey Conservation Area in Idaho during the 1970s).[7] We also look at how newfound city landscapes enable growing populations of raptors to broaden their traditional niches by including urban areas and other human-dominated landscapes in their repertoires of "appropriate" habitats.[8]

Pre-urban Symbiotic Associations between Raptors and Humans

To understand the ecological basis of the phenomenon of urban raptors, it helps to outline the history of symbiotic relations between humans and raptors. Today many hunter-gatherers—including, for example, the Hadza of northern Tanzania[9]—routinely monitor the flights of Old World vultures and follow these avian scavengers to large carcasses that the hunter-gatherers then consume, a behavior that many anthropologists suggest originated millions of years ago when early hominins began doing so across the savannas of Africa's Great Rift

Valley.[2] More recently, pastoralists and transhumant populations (i.e., seasonally moving populations of pastoralists and their herds) "turned the ecological table" on this symbiotic relationship when they began concentrating large flocks and herds of domesticated ungulates that vultures were attracted to and depended on as predictable sources of carrion.[10,11,12,13]

Although it is unknown when raptors first began to live in human settlements, in all likelihood it happened early in our history.[14] Primitive encampments that included refuse almost certainly attracted vultures and other scavenging birds of prey. This would have been especially true for smaller raptors, which were more likely than larger species to have been accommodated and not persecuted by humans.[15]

More than most groups of birds, raptors have captured humanity's imagination for thousands of years.[16] Falconry, an early symbiotic relationship involving raptors and humans, also is associated, albeit indirectly, with the urbanization of raptors. The practice of capturing wild prey using trained raptors dates at least as far back as 4,000 years ago when Asian cultures first began capturing migrating birds of prey and training them to work together with human handlers to capture quarry for the nutritional benefits of both the birds and humans. Although now practiced largely as a sport, falconry flourished as an important hunting technique for humans, particularly in pre-gunpowder days.[16] The art of falconry, by introducing humans to the birds in a positive and nonthreatening way, was an instrumental first step to the more recent urbanization of raptors by making the raptors' presence in human-dominated habitats more likely to be tolerated. Several techniques associated with falconry, including both captive breeding and "hacking" (a process in which nestlings and fledglings are kept and fed for several weeks at hack boards, where food is left for them as they transition to independent hunting), together have allowed conservationists to "soft release" or "hack" young captive-bred peregrine falcons into cities and other landscapes. The hacked young are imprinted on city environments, which has contributed to the growing urban populations of this near cosmopolitan species.[17,18] As a result, by the early 1990s, 34 percent of reintroduced peregrine falcons in the eastern United States were nesting in cities, as were 58 percent of midwestern populations.[19]

The common thread in the early symbiotic relationships involved increasing food availability for raptors, humans, or both. Once humanity began constructing buildings and growing trees agriculturally, the latter both for fuel and building materials, a second element of symbiosis entered the equation: safe nesting sites.

Although built-up areas can be associated with the destruction of natural nesting sites[4] and reduced breeding densities of birds, they can also result in the opposite for raptors.[4] This is particularly the case for smaller and cavity-nesting raptors. Relatively small raptors like bat falcons (*Falco rufigularis*) and Eurasian sparrowhawks (*Accipiter nisus*) pose less of a threat[20,21,22,23] and are less likely to be considered vermin and persecuted by humans.[4] Cavity-nesting raptors like kestrels find that the ledges, holes, nooks, and crannies associated with human architecture are suitable structures in and on which to nest.

Medieval and More Recent Associations between Raptors and Humans

KITES

Red kites (*Milvus milvus*) were said to have "thrived" and nested in London during the reign of King Henry VIII (1509–47)[24] and reportedly were both numerous and protected in Edinburgh in 1600.[25] A foreigner who visited London in the late 15th century would have been "astonished by the enormous number of kites he saw flying round London Bridge."[14] Given the combination of high human densities in larger cities and poor sanitation at the time, there appears to have been plenty of food for these scavengers. Indeed, in mid-16th-century London and Edinburgh, kites often snatched food out of "children's hands on city streets"[14] (figure 1.1), much as black kites (*Milvus migrans*) continue to do in parts of Africa and East Asia,[26,27] where at least until recently, large populations of kites nested in cities.[28] In many such instances, writers remarked about the boldness of the birds and that city inhabitants at the time were quite willing to accept the birds' audacity in light of their value in removing rotting garbage from urban backyards and thoroughfares.[14,26]

FALCONS

Peregrine falcons have long been attracted to cities by the large numbers of rock pigeons (*Columba livia*) and Eurasian starlings (*Sturnus vulgaris*) that inhabit them.[29] In fact, reports suggest that the species has been comfortable in towns and cities since the Middle Ages. Salisbury Cathedral in Wiltshire, England, has hosted nesting peregrine falcons sporadically at least since the mid-1860s.[30] Today, peregrines nest in dozens, if not hundreds, of cathedrals in England.[31] One of the more famous historic North American examples of peregrine falcons

Figure 1.1. Red kites in 16th-century London, from Lea (1909).

nesting on a city skyscraper involves the Sun Life Building in Montreal, Quebec, where from 1936 until 1952, peregrines nested on a ledge that had been "enhanced" with a sandbox that provided a nesting scrape. Along with the phenomenon of nest-site imprinting,[32] recent introductions of fledgling peregrines into cities, coupled with a reduction in pesticide impacts, have bolstered the process of "urbanization" for this species.

In the early part of the 20th century, the Richardson's merlin (*Falco columbarius Richardsonii*), one of the three North American subspecies of the merlin (*F. columbarius*), began to expand the northern limits of its wintering range from Colorado and Wyoming into southwestern Canada. Reports of the expansion document overwintering merlins in Saskatchewan, Canada, in 1922, and in Alberta, Canada, in 1948. The expansion, continuing well into the second half of the 20th century, was especially apparent in urban areas, with Christmas Bird Counts suggesting substantial increases in several of Canada's prairie cities from the late 1950s into the early 1980s. By 1970, Richardson's merlins were not only overwintering in cities but beginning to breed there as well.[33] Since then, nonmigratory populations of "city" merlins have appeared in numerous urban areas throughout southern Canada and the northern United States.

Several factors seem to have played a role in this shift from migratory to nonmigratory behavior. The initial northward expansion of winter areas coincided with the regional expansion of the species' predominant urban prey, the house sparrow (*Passer domesticus*), an Old World species introduced into North America in the 1850s that spread into the American West in the early part of the 20th century.[34] It seems likely that increased prey availability, including both house sparrows and Bohemian waxwings (*Bombycilla garrulus*)—the latter being attracted to urban areas by fruit-bearing ornamental trees—contributed substantially to the merlin's wintering farther north. A second factor was likely the availability of corvid nests in urban areas; merlins, like other falcons, do not construct their own nests but readily use those of other similar-sized birds. Finally, declining human persecution throughout the period may also have played a role by allowing the species to take advantage of this new opportunity.

The lesser kestrel (*Falco naumanni*) also routinely nests in cities and, apparently, has done so for some time, most likely in part because of lower predation on their nestlings.[23] An Old World species that breeds colonially in the architectural nooks and crannies of chapels, churches, and cathedrals, lesser kestrels are aerial insectivores that routinely feed on insects attracted to artificial nighttime lighting at such sites. Detailed observations at well-lit buildings in Seville, Spain, including the city's main cathedral, reveal substantial nighttime hunting by lesser kestrels and nocturnal provisioning of their nestlings during the breeding season. Several of the city's historic buildings are currently illuminated at night for tourists, and the lighting attracts enormous numbers of flying insects, which in turn attract large numbers of aerial insectivores, including both bats and lesser kestrels.[35] The extent to which nocturnal hunting improves the nesting success of these and other kestrels breeding in urban areas has not been studied in detail, but it may be substantial as urban sprawl increases the time it takes parental kestrels to ferry insect prey back to the city from agricultural areas surrounding the city center.[23,36] Peregrines are also attracted to urban lighting by concentrations of disoriented songbirds upon which they feed.[37]

SCAVENGING RAPTORS

In the 1960s, black kites and white-rumped vultures (*Gyps bengalensis*) were nesting at densities of 16 and 2.7 pairs per km², respectively, in Delhi, India. This, together with a smaller population of Egyptian vultures (*Neophron percnopterus*), resulted in an overall urban population estimated at 2,900–3,000 raptors, which were mainly concentrated in "Old Delhi" in mango gardens.[28] The abundance of raptors in Delhi was attributed to three things: (1) food abundance in rubbish heaps, (2) trees for roosting in gardens and along streets, and (3) the "traditional good-will of Indians to all living things" (considered the most important).[28] In Africa, but only north of the equator, the continent's two smallest vultures, the Egyptian vulture and the hooded vulture (*Necrosyrtes monachus*), routinely forage in the immediate vicinity of humans and frequently nest nearby.[10,38,39]

Overwintering adult and nonbreeding juvenile and subadult Egyptian vultures occur in large numbers in refuse dumps associated with the capital of Addis Ababa, a city of more than three million humans in Ethiopia, as well as on migration in southern Israel.[40] The same appears to be true in the Arabian Peninsula around a municipal landfill on the outskirts of Muscat, Oman, a city of 1.5 million.[41] Counts at the Muscat landfill between autumn 2013 and spring 2015 indicate wintertime peaks of between 350 and 450 birds, approximately two-thirds of which were adults, many of which were presumed to be migrants from European breeding populations.[41]

In Uganda, the relationship between humans and hooded vultures, a critically endangered species, has been studied in detail.[39] In the city of Kampala, which in the early 1970s had a human population estimated at 330,000, hooded vultures routinely fed on human rubbish in two large refuse dumps as well as at an abattoir (i.e., slaughterhouse), areas they shared with other avian scavengers, including marabou (*Leptoptilos crumenifer*), black kites, and pied crows (*Corvus albus*). Hooded vultures were particularly common at the abattoir, where their numbers were estimated at more than 100 individuals daily. Numbers there exhibited no consistent seasonal trends, suggesting that the birds using the dump consisted either of nonbreeders or breeders that were nesting nearby. The considerably larger marabous behaviorally dominated the vultures, which in turn behaviorally dominated the smaller pied crows. Black kites sometimes secured food that vultures had consumed by chasing them in the air and forcing them to regurgitate. Although the vultures often perched on buildings in the city, they were not seen nesting on them. It was suspected that the vultures had become numerous in the city because they were "rarely molested unless they became a blatant nuisance," a statement that appears to be true of urban populations of vultures elsewhere in Africa.[39]

Hooded vultures also are human commensals in many urban sites in West Africa, including, at least until recently, Accra, Ghana, a city of more than two million, and Kumasi, Ghana, a city of about two million, where they are not molested.[42] Similarly, both in Banjul, the Gambian capital city of about 35,000, and in exurban areas within 25 km of it, the densities of hooded vultures climbed from 2.9 birds per linear km of road surveyed in 2005 to 12 birds per km in 2013 and to 17.5 birds per km in 2015.[43,44] In western Gambia, hooded vultures are relatively fearless of humans, most of whom value their services in cleaning up humans' rubbish, including moribund fish bycatch, roadkill, butchery scraps at small open-air butcher shops and larger abattoirs, and blood and ruminant stomach contents dumped purposefully for the birds in population centers. Assuming that recent survey results accurately reflect their density,[44] westernmost Gambia hosts a population of 7,000–10,000 individuals, or 4–5 percent of the currently estimated global population of hooded vultures,[45] in an area that represents less than 0.0001 percent of the species range.[44,45]

There is some disagreement among researchers about how long humans and hooded vultures have had a mutually beneficial relationship—some think it is a relatively recent phenomenon, whereas others believe that the "relationship goes back thousands of years."[38] For both hooded and Egyptian vultures, urban commensalism based on food provisioning at refuse dumps and abattoirs is almost certainly enhanced by the absence of larger and competitively superior vultures.[41,42,44] The conservation significance of the sometimes urban nature of the two species may be considerable, as global populations of both vultures are in decline, resulting in their International Union for Conservation of Nature (IUCN) listings as critically endangered (hooded vulture)[46] and endangered (Egyptian vulture).[45,47] Although it has yet to be studied in detail, the commensal nature of the two species will likely play an important role in their survival, at least in the short term.

In southwestern Europe, the Iberian population of griffon vultures (*Gyps fulvus*) has recently begun to grow and increasingly uses urban rubbish dumps, including the one associated with the city of Algeciras.[48] Hundreds of mainly juvenile and subadult griffon vultures feed at the site while waiting for appropriate weather conditions to migrate across the Strait of Gibraltar.[49,50]

In the New World, populations of both black vultures (*Coragyps atratus*) and, to a lesser extent, turkey vultures (*Cathartes aura*) also have been urbanized, particularly in the neotropics. In 1839, Charles Darwin reported that although black vultures were not found near Montevideo, Uruguay, "at the time of the [Spanish] conquest," they had subsequently followed people there.[51] In 1839, John Audubon also mentioned the species' "half domesticated" nature and the ease at which it found food in villages and towns.[52] More recently, black vultures were abundant at a settlement on the Rio Negro in Patagonian Argentina, where they "crowded together in the thousands on trees" at a roost near "cattle slaughtering establishments."[53] In addition, the black vulture extended its range in Brazil considerably "as it accompanie[d] human occupation," and the species routinely breeds on the rooftops of tall buildings in São Paulo and other urban areas.[54,55] Turkey vultures too are sometimes attracted in large numbers to urban areas in southern Central America and northern South America, including Panama City, Panama, particularly in boreal winter when larger migrants arriving from North America successfully compete with relatively smaller tropical black vultures at rubbish tips,[56] inducing "reciprocal migration" of the latter there.[57] Farther south in the Americas, where black vultures currently do not live, nonmigratory subspecies of turkey vultures frequent urban and suburban trash tips in both Tierra del Fuego and the Falkland Islands.[58]

Summary

Continued growth of urban areas occurring around the world[59,60] will have important consequences for diurnal birds of prey, all of which are firmly rooted in fundamental ecological principles. Below we highlight five of them.

1. *As long as urban areas provide "safe bedrooms" for nesting and overlapping or adjacent "well-stocked kitchens" for feeding, breeding populations of at*

least some species of raptors will be attracted to them. This may sometimes occur in large numbers, which may in turn make them less acceptable to people. The latter may be especially true for species with a long history of urbanization. The New World black vulture, a species that has been associated with urban areas for hundreds of years, is a case in point.[51,52] As North American populations of black vultures continue to grow substantially,[61,62] concerns associated with increasing urban populations of the species—including property damage, so-called nuisance roosts, and collisions—continue to arise.[63,64] Aircraft collisions with this species alone are estimated to have cost the US Air Force in excess of $25 million over a 25-year period.[65] In South America, concerns regarding geographically expanding and growing populations of black vultures focus on their apparent effect on Andean condors (*Vultur gryphus*), a globally near-threatened species with a moderately small global population.[45] In some circumstances, Andean condors are outcompeted by black vultures at carcasses.[55] Similar concerns are likely to occur wherever and whenever urban raptors move from being "boutique" members of urban environments to more fully functional, common species in cities.

2. *The likelihood that a raptor will be successful in urban areas will depend on its perceived threat to humans living there.* As a result, relatively small species of raptors are more likely to thrive in urban areas than larger species, particularly if they are asocial or are only moderately social, and do not feed on animals that are viewed as valuable to people. Turkey vultures (but perhaps not black vultures) in the New World are a case in point, as are Egyptian and hooded vultures in the Old World. All of these species are relatively small vultures that, at least historically, appear to have been accepted and even protected by local people.[44,66] Lesser kestrels and peregrine falcons also appear to be notable examples of this principle, in part because of their diets—largely insects in the case of lesser kestrels[35] and feral rock doves in the case of peregrine falcons.[19]

3. *Urban areas that allow an increase in reproductive success, a reduction in mortality, or both can affect regional and even global populations of raptors.* Cases in point include griffon vultures in Spain,[48,49] hooded vultures in parts, though not all, of West Africa,[44] and Egyptian vultures overwintering in North Africa and the Arabian Peninsula.

4. *Abundant, widespread generalist species are more likely to colonize and thrive in urban landscapes than uncommon, limited-range specialist species.*[67] The overwhelming preponderance of examples offered above suggests that relatively widespread species known for their generalist food habits are most likely to be urbanized.

5. *Migratory populations of raptors may be urbanized during certain portions of their annual cycles more than others.* This is especially likely during migration and on the wintering grounds when territoriality is less common. In the New World, both sharp-shinned hawks (*Accipiter striatus*) and Cooper's hawks (*Accipiter cooperii*) are now regular visitors to backyard bird feeders where they prey on "bird feeder" birds.[68] And the migration of at least the former species has changed via migration short-stopping in response to this newfound food source.[69] In the Old World, Egyptian vultures and griffon vultures certainly typify this, given their appearance in large numbers at trash dumps during migration.[41,49] In the New World, turkey vultures do the same in winter,[56] although it should be noted that this species also routinely nests in abandoned and underused buildings in both suburban and urban areas.

Finally, the extent to which the urbanization of wildlife affects evolutionary processes is only now being studied by ecologists and conservation biologists.[67,70] In the long term, we believe that such evolutionary effects may outpace ecological consequences in their overall effect on raptor populations, as well as on raptor diversity and conservation.

Acknowledgments

We thank the family of Sarkis Acopian and the Hawk Mountain Sanctuary Association for supporting our research. Clint Boal and Cheryl Dykstra reviewed and improved our chapter considerably. This is Hawk Mountain Sanctuary's contribution to conservation science number 271.

Literature Cited

1. Oxford University Press. 2002. *Shorter Oxford Dictionary.* 5th ed. Oxford, UK: Oxford University Press.
2. Bickerton, D. 2009. *Adam's Tongue: How Humans Made Language, How Language Made Humans.* New York: Hill and Wang.
3. Ruxton, G. D., and D. M. Wilkinson. 2012. "Endurance Running and Its Relevance to Scavenging by Early Hominins." *Evolution* 67:861–67.
4. Marzluff, J. M., R. Bowman, and R. Donnelly. 2001. *Avian Ecology and Conservation in an Urbanizing World.* Boston: Kluwer Academic.
5. Bird, D. M., D. E. Varland, and J. J. Negro, eds. 1996. *Raptors in Human Landscapes: Adaptations to Built and Cultivated Environments.* San Diego: Academic Press.
6. Cody, M., ed. 1985. *Habitat Selection in Birds.* Orlando: Academic Press.

7. Stuebner, S. 2002. *Cool North Wind: Morley Nelson's Life with Birds of Prey.* Caldwell: Caxton Press.

8. Vandermeer, J. 1972. "Niche Theory." *Annual Review of Evolution and Systematics* 3:107–32.

9. O'Connell, J. F., K. Hawkes, and N. Blurton Jones. 1988. "Hadza Scavenging: Implications of Plio/Pleistocene Hominid Subsistence." *Current Anthropology* 29:356–63.

10. Gavashelishvili, L. 2005. *Vultures of Georgia and the Caucasus.* Tbilisi, Georgia: Georgian Center for the Conservation of Wildlife.

11. Olea, P. P., and P. Mateo-Tomás. 2009. "The Role of Traditional Farming Practices in Ecosystem Conservation: The Case of Transhumance and Vultures." *Biological Conservation* 142:1844–53.

12. Mateo-Tomás, P. 2013. "The Role of Extensive Pastoralism in Vulture Conservation." In *Proceedings of Griffon Vulture Conference*, edited by C. Papazoglu and C. Charalambous, 104–12. Cyprus, Nicosia: BirdLife.

13. Moleón, M. J., A. Sánchez-Zapata, A. Margalida, M. Carrete, N. Owen-Mith, and J. A. Donázar. 2014. "Humans and Scavengers: The Evolution of Interactions and Ecosystem Services." *BioScience* 64:394–403.

14. Lea, J. 1909. *The Romance of Bird Life.* London, UK: Seeley and Co.

15. Bildstein, K. L. 1978. "Behavioral Ecology of Red-Tailed Hawks (*Buteo jamaicensis*), Rough-Legged Hawks (*B. lagopus*), Northern Harriers (*Circus cyaneus*), American Kestrels (*Falco sparverius*), and Other Raptorial Birds Wintering in South-Central Ohio." PhD diss., Ohio State University.

16. Burnham, W. A. 1990. "Raptors and People." In *Birds of Prey*, edited by I. Newton, 170–89, New York: Facts on File.

17. Cade, T. J. 2003. "Starting the Peregrine Fund at Cornell University and Eastern Reintroduction." In *Return of the Peregrine*, edited by T. J. Cade and W. Burnham, 73–104. Boise, ID: The Peregrine Fund.

18. White, C. M., T. J. Cade, and J. H. Enderson. 2013. *Peregrine Falcons of the World.* Barcelona, Spain: Lynx Edicions.

19. Cade, T. J., A. M. Martell, P. Redig, G. Septon, and H. Tordoff. 1996. "Peregrine Falcons in Urban North America." In *Raptors in Human Landscapes: Adaptations to Built and Cultivated Environments*, edited by D. Bird, D. Varland, and J. J. Negro, 3–13. San Diego: Academic Press.

20. Botelho, E. S., and P. C. Arrowood. 1996. "Nesting Success of Western Burrowing Owls in Natural and Human-Altered Environments." In *Raptors in Human Landscapes: Adaptations to Built and Cultivated Environments*, edited by D. Bird, D. Varland, and J. J. Negro, 61–68. San Diego: Academic Press.

21. Bloom, P. H., and M. D. McCrary. 1996. "The Urban Buteo: Red-Shouldered Hawks in Southern California." In *Raptors in Human Landscapes: Adaptations to Built and*

Cultivated Environments, edited by D. Bird, D. Varland, and J. J. Negro, 31–39. San Diego: Academic Press.

22. Parker, J. W. 1996. "Urban Ecology of the Mississippi Kite." In *Raptors in Human Landscapes: Adaptations to Built and Cultivated Environments*, edited by D. Bird, D. Varland, and J. J. Negro, 45–52. San Diego: Academic Press.

23. Tella, J. L., F. Hiraldo, J. A. Donázar-Sancho, and J. J. Negro. 1996. "Costs and Benefits of Urban Nesting in the Lesser Kestrel." In *Raptors in Human Landscapes: Adaptations to Built and Cultivated Environments*, edited by D. Bird, D. Varland, and J. J. Negro, 53–60. San Diego: Academic Press.

24. Ratcliffe, D. 1997. *The Raven*. San Diego: Academic Press.

25. Nash, J. K. 1935. *The Birds of Midlothian*. London, UK: Witherby.

26. Brown, L., and D. Amadon. 1968. *Eagles, Hawks and Falcons of the World*. New York: McGraw-Hill.

27. Meinertzhagen, R. 1959. *Pirates and Predators: The Piratical and Predatory Habits of Birds*. Edinburgh, Scotland: Oliver and Boyd.

28. Galushin, V. M. 1971. "A Huge Population of Birds of Prey in Delhi, India." *Ibis* 113:522.

29. Ratcliffe, D. 1993. *The Peregrine Falcon*. 2nd ed. Carlton, UK: T. and A. D. Poyser.

30. Bildstein, K. L. 2017. *Raptors: The Curious Nature of Diurnal Birds of Prey*. Ithaca: Cornell University Press.

31. Drewitt, E. 2014. *Urban Peregrines*. Exeter, UK: Pelagic Publishing.

32. Cade, T. J., and W. Burnham, eds. 2003. *Return of the Peregrine*. Boise, ID: The Peregrine Fund.

33. Warkentin, I. G., P. C. James, and L. W. Oliphant. 1990. "Body Morphometrics, Age Structure, and Partial Migration of Urban Merlins." *Auk* 107:25–34.

34. Sodhi, N. S., and L. W. Oliphant. 1993. "Prey Selection by Urban-Breeding Merlins." *Auk* 110:727–35.

35. Negro, J. J., J. Bustamante, C. Melguizo, J. L. Ruiz, and J. M. Grande. 2000. "Nocturnal Activity of Lesser Kestrels under Artificial Lighting Conditions in Seville, Spain." *Journal of Raptor Research* 34:327–29.

36. Liven-Schulman, I., Y. Leshem, D. Alon, and Y. Yom-Tov. 2004. "Causes of Population Declines of the Lesser Kestrel *Falco naumanni* in Israel." *Ibis* 146:145–52.

37. DeCandido, R., and D. Allen. 2006. "Nocturnal Hunting by Peregrine Falcons at the Empire State Building, New York City." *Wilson Journal of Ornithology* 118:53–58.

38. Mundy, P., D. Butchart, J. Ledger, and S. Piper. 1992. *The Vultures of Africa*. London, UK: Academic Press.

39. Pomeroy, D. E. 1975. "Birds as Scavengers of Refuse in Uganda." *Ibis* 117:69–81.

40. Yosef, R. 1996. "Raptors Feeding on Migration in Eilat, Israel: Opportunistic Feeding or Migration Strategy." *Journal of Raptor Research* 30:242–45.

41. Al Fazari, W. A., and M. J. McGrady. 2016. "Counts of Egyptian Vultures *Neophron percnopterus* and Other Avian Scavengers at Muscat's Municipal Landfill, Oman, November 2013–March 2015." *Sandgrouse* 38:99–105.

42. Campbell, M. 2009. "Factors for the Presence of Avian Scavengers in Accra and Kumasi, Ghana." *Area* 41:341–49.

43. Barlow, C. R., and T. Fulford. 2013. "Road Counts of Hooded Vultures *Necrosyrtes monachus* over Seven Months in and around Banjul, Coastal Gambia, in 2005." *Malimbus* 35:50–56.

44. Jallow, M., C. R. Barlow, L. Sanyang, L. Dibba, C. Kendall, M. Bechard, and K. L. Bildstein. 2016. "High Population Density of Critically Endangered Hooded Vulture *Necrosyrtes monachus* in Western Region, The Gambia, Confirmed by Road Surveys in 2013 and 2015." *Malimbus* 38:23–28.

45. BirdLife International. 2016. "Species Factsheets." Accessed October 6, 2016. http://birdlife.org.

46. Ogada, D. L., and R. Buij. 2001. "Large Declines of the Hooded Vulture *Necrosyrtes monachus* across Its African Range." *Ostrich* 82:101–13.

47. Velevski, M., S. C. Nikolov, B. Hallmann, V. Dobrev, L. Sidiropoulos, V. Saravia, R. Tsiakiris, V. Arkumarev, A. Galanaki, T. Kominos, K. Stara, E. Kret, B. Grubac, E. Lisicanec, T. Kastritis, D. Vavylis, M. Topi, B. Hoxha, and S. Oppel. 2015. "Population Decline and Range Contraction of the Egyptian Vulture *Neophron percnopterus* in the Balkan Peninsula." *Bird Conservation International* 25:440–50.

48. Parra, J., and J. L. Tellería. 2004. "The Increase in the Spanish Population of Griffon Vultures *Gyps fulvus* during 1989–1999: Effects of Food and Nest-Site Availability." *Bird Conservation International* 14:33–41.

49. Garrido, J. R., C. G. Sarasa, and M. Fernández-Cruz. 2002. "Rubbish Dumps as Key Habitats for Migration and Wintering in the Griffon Vulture (*Gyps fulvus*) at a Migratory Bottleneck: Implications for Conservation." In *Raptors in the New Millennium*, edited by R. Yosef, M. L Miller, and D. Pepler, 143–51. Eilat, Israel: The International Birding and Research Center.

50. Bildstein, K. L., M. J. Bechard, C. Farmer, and L. Newcomb. 2009. "Narrow Sea Crossings Present Major Obstacles to Migrating Griffon Vultures *Gyps fulvus*." *Ibis* 151:382–91.

51. Darwin, C. 1839. *Voyage of the Beagle*. London, UK: Henry Colburn.

52. Audubon, J. J. 1839. *The Birds of America*. London, UK: Constable and Company.

53. Hudson, W. H. 1920. *Birds of La Plata*. Vol. 2. London, UK: J. M. Dent and Sons.

54. Sick, H. 1993. *Birds in Brazil*. Princeton: Princeton University Press.

55. Carrete, M., S. A. Lambertucci, K. Speziale, O. Ceballos, A. Travaini, M. Delibes, F. Hiraldo, and J. A. Donázar. 2010. "Winners and Losers in Human-Made Habitats: Interspecies Competition Outcomes in Two Neotropical Vultures." *Animal Conservation* 13:390–98.

56. Koester, F. 1982. "Observations on Migratory Turkey Vultures and Lesser Yellow-Headed Vultures in Northern Colombia." *Auk* 99:372–75.
57. Bildstein, K. L., M. J. Bechard, P. Porra, E. Campo, and C. J. Farmer. 2006. "Seasonal Abundances and Distributions of Black Vultures (*Coragyps atratus*) and Turkey Vultures (*Cathartes aura*) in Costa Rica and Panama: Evidence for Reciprocal Migration in the Neotropics." In *Neotropical Raptors*, K. L. Bildstein, D. R. Barber, and A. Zimmerman, 47–60. Orwigsburg, PA: Hawk Mountain Sanctuary.
58. Augé, A. 2016. "Anthropogenic Debris in the Diet of Turkey Vultures (*Cathartes aura*) in a Remote and Low-Populated South Atlantic Island." *Polar Biology* 40:799–805.
59. Wigginton, N. S., J. Fahrenkamp-Uppenbrink, B. Wible, and D. Malakoff. 2016. "Cities Are the Future." *Science* 352:904–7.
60. Liu, Z., C. He, Y. Zhou, and Wu. 2014. "How Much of the World's Land Has Been Urbanized, Really? A Hierarchical Framework for Avoiding Confusion." *Landscape Ecology* 29:763–71.
61. Buckley, N. J. 1999. "Black Vulture (*Coragyps atratus*)." In *The Birds of North America*, edited by P. G. Rodewald. Ithaca: Cornell Lab of Ornithology. Accessed February 6, 2017. https://birdsna.org/Species-Account/bna/species/blkvul.
62. Avery, M. L. 2004. "Trends in North American Vulture Populations." *Proceedings of the Vertebrate Pest Conference* 21:116–21.
63. Lowney, M. 1999. "Damage by Black and Turkey Vultures in Virginia, 1990–1996." *Wildlife Society Bulletin* 27:715–19.
64. Avery, M. L., and J. L. Cummings. 2004. "Livestock Depredations by Black Vultures and Golden Eagles." *Sheep and Goat Research Journal* 19:58–63.
65. US Air Force (USAF). 2015. "Factsheet." Air Force Safety Center. Accessed October 14, 2016. http://www.safety.af.mil/Divisions/Aviation-Safety-Division/BASH.
66. Wilson, A. 1840. *Wilson's American Ornithology*. Boston: Otis, Broaders, and Company.
67. Evans, K. L., D. E. Chamberlain, B. J. Hatchwell, R. D. Gregory, and K. J. Gaston. 2011. "What Makes an Urban Bird?" *Global Change Biology* 17:32–44.
68. Roth, T. C., II, W. E. Vetter, and S. L. Lima. 2008. "Spatial Ecology of Wintering *Accipiter* Hawks: Home Range, Habitat Use, and the Influence of Bird Feeders." *Condor* 110:260–68.
69. Viverette, C. B., S. Struve, L. J. Goodrich, and K. L. Bildstein. 1996. "Decrease in Migrating Sharp-Shinned Hawks (*Accipiter striatus*) at Traditional Raptor-Migration Watch Sites in Eastern North America." *Auk* 113:32–40.
70. Møller, A. P. 2010. "Interspecific Variation in Fear Responses Predicts Urbanization in Birds." *Behavioral Ecology* 21:365–71.

CHAPTER 2

City Lifestyles:
Behavioral Ecology of Urban Raptors

Cheryl R. Dykstra

IF THERE IS A SINGLE UNIFYING CHARACTERISTIC of urban/suburban wildlife species, it is likely adaptability. Species that can occupy urban areas are behaviorally flexible,[1,2] and this flexibility drives changes in the way they use urban space and cohabit with people. Raptors too exhibit behavioral changes when they move from rural to urban environs.[3] Inherent plasticity allows some raptors to adjust their behavior to survive in circumstances that may differ greatly from those of more typical, rural, or natural areas. For example, they may move into urban environments that are suitable for them,[4,5] or they may persist by tolerating human activity in a natural area that has been overwhelmed by suburbia. They may perceive human-made objects such as rooftops, utility towers, billboards, and bridges as potential nest sites,[3,6,7,8] especially in areas where nest sites in traditional, natural locations are limited. Additionally, they may take advantage of a different prey type that is present, such as rats (*Rattus* spp.),[9,10] or a typical prey type that is more abundant or available (e.g., birds at feeders).[5,11]

Raptor species' differing adaptabilities, in combination with the variety of urban environments within and among cities, have led to great diversity in raptor urban ecologies. Different species, and even different populations of the same species, exhibit varying degrees of adaptability[3] and are not equally flexible in all aspects of their behavior (e.g., urban screech-owls [*Megascops asio*] shift to eating more avian prey but still require a cavity for nesting).[12] Likewise, urban habitat is fragmented and diverse, with multiple influences that vary widely (e.g., pollution, prey base, predators, nest-site availability). Thus it is difficult to make generalizations about urban raptor ecology and behavior. However, that very diversity also makes urban raptors incredibly interesting to study, and as seen in chapters 6–13, the scientists studying them are passionate about understanding our wild raptor neighbors and helping local human residents better appreciate them.

Differences in raptor behavior in urban areas sometimes promote variation in demographic traits, such as reproductive rate and survival (see chapter 4). For example, feeding on smaller, low-energy prey or reducing attendance at the nest because of disturbance or a need to range farther to find prey can result in reduced reproductive success compared to birds in more typical rural environments.[13] Conversely, hunting at bird feeders with abundant prey may lead to higher reproductive rates.[13] Foraging in the short vegetation alongside roadways can lead to increased mortality due to collisions with vehicles, and selecting nesting sites on highline transmission towers may lead to increased risk of electrocution.[14] Behavior thus underlies demography and population ecology, and differences in behavior between urban and rural birds can illuminate the effects of urbanization on raptors, giving insight into how raptors coexist with humans in urban and suburban spaces.

Habitat Used, Nest-Site Selection, and Nests Used

Urban and suburban landscapes are fragmented mosaics of industrial, residential, and recreational areas and patches of native vegetation. Raptors inhabiting these landscapes select a home range, a nest site, and foraging areas, and these selections may vary depending on season, year, and even time of day. In addition, habitat conversion may force raptors to shift the areas they use, because urban development can occur rapidly, especially on the suburban fringe of cities. Studies of habitat use can provide valuable data for assessing the suitability of landscapes for raptors and estimating potential size of urban raptor populations.

However, the comparison of urban areas among studies is complicated by diverse and varied definitions of land-cover categories. *Low-density residential* in one study may equate to *residential* or even *high-density residential* in another study, or such terms may be undefined. Fragmented landscapes described using remote-sensing imagery may be classified differently when using 30 m² pixels (parcels) versus 1 m² pixels. Even the classification of landscapes as *urban, suburban,* and *exurban* is not standardized. To simplify, I here use *urban* for all land-cover types within cities and their suburbs (except where an author has differentiated urban from suburban) and *rural* for all land-cover types in more natural areas with low human population density.

Interpretation of urban raptor behavior and habitat use is hampered by the limited number of studies with a complete urban versus rural study design. To make ideal comparisons, researchers should measure behaviors in both urban areas and nearby rural areas. This would ensure methodologies are consistent and local factors affecting the species' ecology (weather, latitude, general native habitat, native prey base, and predators) are similar between the sites or at least can be recognized and quantified. More commonly, researchers cannot undertake these complete studies, and instead they must qualitatively compare the urban behaviors they measure to data for the same species reported by other investigators in more typical rural areas; direct comparisons between these studies may be difficult due to differences in measuring and reporting the abundances of different types of land covers and other variables and differences in the species' ecology when compared populations are far apart in space or time.

Habitats Used by Urban Raptors

Urban raptors use a wide variety of habitats, ranging from highly urban city centers[15,16] to low-density residential housing interspersed with pockets of native vegetation,[17,18,19] and quantifications of habitat within home ranges reflect that diversity (table 2.1). Even within a single study, individual birds' habitats can differ widely.[20]

The environments normally used by urban raptors often differ from those available, with some species avoiding human-use areas and others selecting them. Breeding Cooper's hawks (*Accipiter cooperii*) used parks or ornamental plantings and commercial or industrial areas disproportionately more than expected in southern California,[21] and residential areas and parks or golf courses were used more in Tucson, Arizona.[20,22] Similarly, wintering Cooper's hawks used residential areas disproportionately in Indiana.[23]

On the other hand, long-eared owls (*Asio otus*), great horned owls (*Bubo virginianus*), and red-shouldered hawks (*Buteo lineatus*) avoided areas with human activity. The owls selected wooded areas and meadows[24] or native forests and grasslands,[25] and the hawks chose pond edges and riparian zones.[26]

Urban northern goshawks (*Accipiter gentilis*) nested in parks in Hamburg, Germany, and spent 88 percent of their daytime hours there but made foraging

Table 2.1. Home-range size and habitat in home ranges for selected urban raptors.

Species	Location	Years	n	Description of study sites	Season	Mean home-range size(s) (ha)	P	Habitat in urban/suburban home ranges	Reference
Cooper's hawk	Orange County, CA	2001	8	Urban vs. natural	breeding	481 (urban) vs. 609 (natural)	NS (confidence intervals)	>60% urban	21
Long-eared owl	Czech Republic	2004–6	9	Urban vs. suburban	breeding	446 (urban) vs. 56 (suburban)	0.02	41% developed area[a]	24
Ferruginous hawk	Denver, CO	1992–95	38	Suburban vs. rural	winter	2,300 (suburban) vs. 4,710 (rural)	0.28		88
Cooper's hawk	Terre Haute, IN	1999–2004	6 adults	Urban vs. rural	winter	390 (urban) vs. 1,420 (rural)	0.06		23
Cooper's hawk	Tucson, AZ	2009–10	23	Urban	breeding	73		60% low-density residential	22
Cooper's hawk	Tucson, AZ	1996–97	9 males	Urban	breeding	66		52% high-density residential	20
Northern goshawk	Hamburg, Germany	1997–99	3 males	Urban	breeding	863		71% built-up	27
Merlin	Saskatoon, SK	1987–90	27	Urban	breeding	1,110		34%–88% urban	28
Red-shouldered hawk	Cincinnati suburbs, OH	1998–2000	11	Suburban	breeding	90		50% suburban; 41% forest[b]	26
Red-tailed hawk	Hartford, CT	2007–10	11	Urban and suburban	annual	108		50% green space	31
Great horned owl	Orange County, CA	1997–98	10[c]	Urban and suburban	annual	348		57% urban	25
Barred owl	Charlotte, NC	2002	9	Suburban	breeding	90		81% residential land use[d]	30

a. Urban owls only.
b. Habitat percentages reported are for annual ranges; breeding and nonbreeding ranges had similar proportions of habitat types.
c. Includes only owls with urban habitat in home ranges.
d. Zoning classification, not ecological.

forays into surrounding built-up areas, where they caught 42 percent of their prey (primarily rock pigeons [*Columba livia*]).[27] Other urban species did not seem to prefer or avoid any particular environments but instead spent time in the various habitat types in direct proportion to their availability.[28] Differences among species and studies are likely a consequence of raptors preferentially seeking out areas with abundant prey, where their morphologies allow them to hunt successfully,[22,24] or areas with suitable nest sites.[15,29]

Home ranges of urban raptors tend to be smaller than, or similar in size to, those used by their rural counterparts, although only a few studies specifically documented urban and rural home ranges simultaneously. Where researchers did measure both urban and rural home ranges, they found that Cooper's hawks in urban and natural areas of southern California used similar-sized home ranges, as did wintering ferruginous hawks (*Buteo regalis*) in Colorado (table 2.1). Wintering Cooper's hawks in urban areas of Indiana used smaller ranges than rural birds.

In urban raptor studies where rural home ranges were not measured for comparison, some authors compared their results to previously published information; although these comparisons lack the ideal urban versus rural study design, they nonetheless reveal the same trends—smaller or similar-size home ranges. For example, red-shouldered hawks, red-tailed hawks (*Buteo jamaicensis*), northern goshawks, Cooper's hawks, and barred owls (*Strix varia*) in urban settings used home ranges that were smaller than those of rural conspecifics in other studies (table 2.1).[20,22,26,27,30,31] Urban great horned owl ranges were judged similar in size to their rural counterparts.[25] Urban long-eared owls had larger home ranges than suburban birds, which was attributed to large amounts of unusable area incorporated into the urban home ranges (table 2.1), but the researchers considered that the urban ranges were similar in size to those of rural conspecifics in other studies.[24]

Sometimes raptors in the same city can have very different space-use patterns, which may be unsurprising given the behavioral flexibility of urban birds. For example, researchers classified merlins (*Falco columbarius*) nesting in the city of Saskatoon into two groups: (1) resident merlins, hatched in the city, which rarely hunted outside the city and had smaller home ranges (630 ha for males), and (2) immigrants, hatched outside the city, which had larger home ranges encompassing rural hunting areas (3,370 ha for males).[28]

Many factors can influence home-range size, including age, sex, and conspecific density, but the most important determinant is likely food availability.[32,33] The observation that urban home ranges are often smaller than, or at least similar to, typical exurban or rural home ranges suggests that prey availability in many urban areas is sufficient or abundant.

Nest-Site Selection

Selection, used here as typically defined by ecologists (i.e., a difference between the areas/environments used [habitat] and available environments) is commonly measured at raptor nests by comparing nest plots to similar plots randomly placed throughout the study area of interest. Plots typically reflect the immediate nest area (0.04–1 ha) or the landscape scale (300–750 ha). In some studies, researchers measured selection at both urban and rural sites, and also compared urban and rural sites to each other.

Nest-site selection among urban raptors varies, but in general, raptors (specifically urban Mississippi kites [*Ictinia mississippiensis*],[34] urban Cooper's hawks,[35] suburban and rural red-shouldered hawks) selected nest trees that were taller and larger (diameter at breast height [DBH]) than central trees in the random plots and had larger, taller trees in the plot surrounding the nest compared to the random plots.[36] However, nest trees and sites of suburban and rural barred owls did not differ from random plots in any measured characteristics, possibly due to small samples sizes.[30]

Urban nest sites of Mississippi kites and red-shouldered hawks were no farther from houses than random plots, indicating that they did not avoid nesting near houses.[34,36] In contrast, urban Cooper's hawks selected nest sites farther from houses and roads than random sites, although nests were still relatively close to human structures, averaging only 42 m from the nearest building.[35]

Unsurprisingly, urban raptors' nesting areas were closer to houses and roads than their counterparts' in rural areas.[30,36,37] Other than these measures, suburban nest sites of some species, such as red-shouldered hawks, differed little from rural ones.[36] But in other species, the nest trees and nesting areas differed in interesting ways. Suburban barred owls' nest trees were larger and nest plots were more open, from the canopy layer down to the ground cover.[30] Suburban red-tailed hawks' nest trees were taller than urban and rural nest trees, and suburban plots around nests had more species of shrubs and more saplings than urban and rural sites.[37]

On a landscape scale, urban raptors tend to select areas with more natural land-cover types. For example, urban and suburban red-tailed hawks avoided the areas of densest urban cover and selected areas with more grassland, more forest, and greater land-cover diversity compared to unused plots.[38] Urban areas occupied by nesting barn owls (*Tyto alba*) in Rome contained more open land and less developed area than random plots, but less deciduous and coniferous forest than rural sites.[39]

Urban nest sites varied widely among studies where nest sites were described quantitatively but were not specifically compared to rural ones in the same study. Nests of urban sharp-shinned hawks (*Accipiter striatus*) in Quebec were situated in conifers within mixed deciduous-conifer stands that had high canopy closure, and were located on average 20 m from an opening that in many cases constituted considerable human activity; researchers noted that the large variability in most measurements suggested significant flexibility in nest-site requirements.[40] Swainson's hawks (*Buteo swainsoni*) in California nested in conifers in residential neighborhoods that were more than 20 years old and preferred neighborhoods older than 45 years old, which provided suitably mature trees for nesting; in the middle-aged neighborhoods (20–45 years old), hawks selected larger-than-average trees.[29] Urban eastern screech-owls more often used, and had higher reproductive rates in, nest boxes that were situated closer to the nearest house, possibly because such sites harbored fewer predators.[12]

Nests Used

Some urban raptor species choose novel nest sites and substrates in cities, whereas others build nests in trees that differ little from those used in rural areas. It is intriguing that different species' willingness to use unusual sites seems roughly aligned with taxonomic classification.

Falcons

In general, falcon species appear most likely to adopt unfamiliar nesting substrates in the urban landscape. Urban peregrine falcons (*Falco peregrinus*) in eastern North America nest on skyscraper ledges, power plant cooling towers, bridges, and in open boxes specifically built for them;[6,16,41] however, because peregrines were reintroduced from hack boxes in cities, their progeny's use of such urban nest sites is perhaps not entirely natural. Other falcons also adapted well to urban sites, with many species using novel sites for their nest scrapes. Eurasian kestrels (*Falco tinnunculus*) nest primarily in flowerpots on windowsills in Israel[42] and in cavities in historic buildings in Vienna.[15] In Spain, researchers established a rooftop colony for lesser kestrels (*Falco naumanni*) in Seville, primarily to facilitate research,[43] while in another study, most urban nests were located in buildings.[44] In contrast to all of these, urban merlins are similar to their rural counterparts by nesting primarily in stick nests built by corvids in conifer trees.[45]

Ospreys

Ospreys (*Pandion haliaetus*) nest in urban areas with sufficient water and are well known for their propensity to use nest platforms and other human-made substrates throughout their ranges.[46,47] Ospreys have also nested in urban areas atop various elevated objects, including a crane, the mast of a sailboat,[46] and an automobile raised 30 m in the air at a car dealership.[46]

Buteos

Buteo species that use urban or suburban areas can be considered moderately flexible in their choice of nest substrate. Urban red-tailed hawks nested in native trees in Milwaukee, Wisconsin, but also on billboards, high-voltage transmission towers, civil defense sirens, and other human-made towers (5.5 percent of nesting attempts).[8] Urban red-shouldered hawks in southern and central California nested frequently in nonnative *Eucalyptus* trees,[48,49] whereas those in Cincinnati, Ohio, primarily used native trees similar in size and height to those selected by their rural counterparts in south-central Ohio.[36] However, human-made structures may also be used; two pairs nested on rooftops, and one pair nested on a gas grill.[7,26] Most (four of five) suburban Swainson's hawks in Saskatchewan nested in trees, but one pair built its nest on a railway signal gantry.[50]

Accipiters

Accipiter species may be least flexible in their choice of nesting substrate. Despite their abundance in urban and suburban North America, Cooper's hawks have not initiated nesting on human-made substrates, although one pair used a nest platform erected in a tree with a pre-built stick nest already on it (R. Rosenfield, pers. comm.), and they will build in nonnative tree species.[22] Likewise, northern goshawks living in Berlin, Hamburg, and other cities in Europe apparently build stick nests only in trees.[4,51]

Owls

Many species of owls readily nest in nest boxes, and urban owls are no exception. Barred owls, tawny owls (*Strix aluco*), eastern screech-owls, and barn owls all used nest boxes, to the delight of the landowners who installed them; at least one such landowner live-streamed video from his barred owl nest box directly to his giant-screen television. Suburban great horned owls accept human-made nesting containers created from one-third of a 55-gallon drum lined with wood chips.[52] Three barred owls in Charlotte, North Carolina, nested inside chimneys.[30]

These and other urban owl species also use natural cavities, snags, and open stick nests appropriated from other raptors, as they do in rural ecosystems. As we will see in chapter 11, lack of cavities in urban areas may be limiting some owls;[30,53] the perceived or real danger to residents and buildings often leads urban landowners to remove trees and limbs that develop cavities.

Foraging and Diet

Urban development significantly alters the populations of many small animal species, and these effects are transferred up the food chain to the top predators, including the urban raptors. Populations of some generalist, disturbance-tolerant prey species—including "pests" such as rats and rock pigeons—increase in urban areas, whereas populations of many other prey species decrease or disappear.[3,54] Populations of "desirable" prey species, such as the birds that utilize bird feeders, may also increase with development.[5,54] Predictably, raptor species' responses to these diverse ecological changes vary widely.

As with the habitat studies, few researchers have compared diets of urban raptors with those in nearby rural environments during the same time period. This reduces our ability to assess how raptors adapt to urban ecosystems and how diet flexibility influences other aspects of the species' ecology, such as reproductive rates (discussed in detail in chapter 4). For studies without such comparisons, simple descriptions of the prey species eaten in urban areas can still provide a base for understanding raptor roles in the urban trophic web.

Increased Prey Abundance or Availability

When compared to raptors in rural areas, many urban raptors consume more synanthropic prey species, including some considered pests. Barn owls, barred owls, and great horned owls in more urban areas of British Columbia ate more rats (*Rattus norvegicus*) than those farther from development,[9,10,55] and barn owls in Argentina ate more rats (*Rattus* spp.) and mice (*Mus musculus*) as urbanization increased along a rural-urban gradient.[56] City peregrines are admired for their tendency to capture feral rock pigeons.[57,58] Others raptors too focus on rock pigeons (northern goshawk)[59] or house sparrows (*Passer domesticus*; tawny owl,[60] Eurasian sparrowhawk [*Accipiter nisus*],[61,62] merlin,[45] Eurasian kestrel).[63]

For avivores, cities can provide an abundance of avian prey.[3] In some cases, avian populations are higher than they are in surrounding natural areas because of the presence of bird feeders or more habitat. As we shall see in more detail in chapter 4, the latter is particularly true for urban areas surrounded by desert, where the urban environment contains vegetation planted and watered for human benefit.[13,64] Cooper's hawks in Tucson, Arizona, delivered more prey to their nestlings than their rural counterparts, and all were birds, including doves, which made up 57 percent of urban prey deliveries.[64] Generalist raptors may also shift their diets toward birds;[11] for example, avian prey made up 55 percent of the diet of suburban barred owls, compared to only 5 percent of that of nearby rural owls.[18]

Increased avian prey base in cities,[65] associated in part with bird-feeding by humans, has been credited for colonization of cities by northern goshawks and Eurasian sparrowhawks[4,27,66] and increased reproductive success in cities compared to rural habitat.[13,51] However, urban Eurasian kestrels ate more birds than their rural counterparts but suffered greater nest failure due to nestling starvation,[15] and bird-eating Cooper's hawks in Tucson, Arizona, experienced higher rates of nest failure due to the parasite *Trichomonas*[65] compared to rural birds. Thus the presence of enhanced bird populations due to bird feeders may attract raptors to nest in urban locations but does not always guarantee success there, leading some researchers to suggest that some urban areas may be ecological traps.[15]

Decreased Prey Abundance or Availability

Some urban environments apparently provide less food for the urban raptors that make their homes there, resulting in lower reproductive rates. Fewer mammals in the diet of nestling Eurasian kestrels in the city center of Vienna, Austria, resulted in higher rates of starvation compared to rural nests outside the city.[15] Similarly, nestlings in an urban population of lesser kestrels in Spain starved to death more often than did rural nestlings.[44]

Increased Scavenging Opportunities

Some raptors species have adapted their foraging behavior to make use of food scavenged (or hunted) in waste disposal areas within urban environs. In Madrid and Rome, black kites (*Milvus migrans*) frequent rubbish dumps,[67,68] and in Japan, they congregate in urban parks to catch food children toss into the air for them (C. Boal, pers. comm.). In Delhi, India, scavenged meat is an important component of the diet of black kites, along with rats and rock pigeons.[69]

TYPICAL DIET

Some urban raptors catch prey that is typical for their species and does not differ from their diet in rural areas. Eurasian kestrels ate mostly voles all along an urban-rural habitat gradient.[70] Burrowing owls (*Athene cunicularia*) in urban areas and rural areas of the Argentine Pampas consumed mostly arthropods (95 percent in urban areas, 95 percent in agroecosystems, 86 percent in vegetated dunes), and the proportions of each taxa were similar; the authors concluded that urban owls do not need alternate food sources to inhabit urban areas.[71] In many studies, diet was quantified only in the urban area, so direct comparison with diet in a nearby rural area was precluded. However, authors indicated that the "typical" prey were caught by urban red-shouldered hawks,[72] Harris's hawks (*Parabuteo unicinctus*),[73] and white-tailed kites (*Elanus leucurus*).[74]

Other Behaviors

Urban raptors' behavioral flexibility[3,5,75,76] is demonstrated by their ability to tolerate the close proximity of humans.[1] Researchers have noted qualitatively that urban raptors are more easily approached, facilitating study such as observations or reading color bands.[19,21,51,65] Urban birds in general have lower flight initiation distance (FID; i.e., the distance at which a perched bird flies away from an approaching researcher)[5,76,77] than rural birds. Researchers have rarely quantified FID of urban raptors; however, urban burrowing owls[78] and Mississippi kites[34] have lower FID than rural ones, and urban, but not rural, burrowing owls perceive a human with a dog as more dangerous than a human alone (i.e., more aggressive behaviors when exposed to the dog and human).[78]

Urban raptors occasionally defend their nests vigorously, diving at city residents near their nests and sometimes hitting them.[21,48,65,79] Urban Mississippi kites are notoriously defensive,[34,79] but Cooper's hawks[21,65,80] and red-shouldered hawks[19,48] also sometimes strike people walking near the birds' nests. Such behaviors can lead to injury and can provoke negative perceptions and responses by human residents, as we shall see in chapter 15. However, many landowners tolerate infrequent hawk attacks with good humor and empathy, particularly if educated by researchers.[19,65]

Behavioral flexibility may allow the development of alternate reproductive strategies in urban areas. For example, polygyny occurs rarely in some urban populations (Cooper's hawk, lesser kestrel).[81,82] Extra-pair copulations (EPC) are frequent in urban Cooper's hawks in Milwaukee, Wisconsin, where 34 percent of urban broods contain at least one young of extra-pair paternity,[83] and very common in colonial-nesting lesser kestrels, in which 85 percent of observed pairs had extra-pair copulations by one or both members of the pair.[84] The frequencies of these behaviors may represent adaptations to urban conditions, such as a dense nesting population or high food abundance;[83] however, it is not clear whether these rates are unusual due to the dearth of comparison data from rural populations. More research is needed to better understand the factors that influence these behaviors.

Time-activity budgets of urban birds also have received little attention, although one might expect that urban birds would alter their activities in response to disturbance by humans, pets, vehicular traffic, and noise.[54] Chipman and colleagues[85] reported that time-activity budgets of urban burrowing owls did not differ from those of rural ones; birds at both locations spent most of their time perching, resting, and being vigilant. Behavioral responses to the urban environment vary among raptors species. In Spain, Spanish imperial eagles (*Aquila adalberti*) and two species of vultures avoided roadsides near a large city during weekends, when traffic volume was heavy, although six other species did not; researchers speculated that vehicle noise or an alteration in prey availability influenced raptor activity cycles.[86]

Concluding Remarks

Behavioral flexibility is a defining characteristic of raptors that can successfully inhabit urban areas. Accordingly, urban raptors vary widely in habitat and nest-site selection, foraging habits, and other behaviors, and they make use of many zones along the urban-to-rural gradient of human-altered habitats. Human disturbance, broadly defined to include landscape and ecosystem modifications, acts as a selecting force in urban areas and shapes human-raptor interactions. This selection pressure may favor species that are flexible and tolerant or tame[5] or may favor tame individuals of species that exhibit variability in behavior, specifically tolerance of humans.[76,77,87] The study of behavior in urban raptors, particularly in comparison to their rural counterparts, can provide insight into the process of urbanization as well as the coevolution of urban wildlife communities.

Literature Cited

1. Luniak, M. 2004. "Synurbization—Adaptation of Animal Wildlife to Urban Development." In *Proceedings of the 4th International Urban Wildlife Symposium*, edited by W. W. Shaw, K. L. Harris, and L. Van Druff, 50–55. Tucson: University of Arizona.

2. Marzluff, J. M. 2016. "A Decadal Review of Urban Ornithology and a Prospectus for the Future." *Ibis* 159:1–13.

3. Chace, J. F., and J. J. Walsh. 2006. "Urban Effects on Native Avifauna: A Review." *Landscape and Urban Planning* 74:46–69.

4. Rutz, C. 2008. "The Establishment of an Urban Bird Population." *Journal of Animal Ecology* 77:1008–19.

5. Møller, A. P. 2009. "Successful City Dwellers: A Comparative Study of the Ecological Characteristics of Urban Birds in the Western Palearctic." *Oecologia* 159:849–58.

6. Septon, G. A., J. Bielefeldt, T. Ellestad, J. B. Marks, and R. N. Rosenfield. 1996. "Peregrine Falcons: Power Plant Nest Structures and Shoreline Movements." In *Raptors in Human Landscapes: Adaptations to Built and Cultivated Environments*, edited by D. Bird, D. Varland, and J. J. Negro, 145–53. San Diego: Academic Press.

7. Hays, J. L. 2000. "Red-Shouldered Hawks Nesting on Human-Made Structures in Southwest Ohio." In *Raptors at Risk: Proceedings of the V World Conference on Birds of Prey and Owls*, edited by R. D. Chancellor and B.-U. Meyburg, 469–71. Berlin, Germany: World Working Group on Birds of Prey and Owls; Surrey, BC: Hancock House Publishers.

8. Stout, W. E., S. A. Temple, and J. M. Papp. 2006. "Landscape Correlates of Reproductive Success for an Urban-Suburban Red-Tailed Hawk Population." *Journal of Wildlife Management* 70:989–97.

9. Hindmarch, S., and J. E. Elliott. 2015. "A Specialist in the City: The Diet of Barn Owls along a Rural to Urban Gradient." *Urban Ecosystems* 18:477–88.

10. Hindmarch, S., and J. E. Elliott. 2015. "When Owls Go to Town: The Diet of Urban Barred Owls." *Journal of Raptor Research* 49:66–74.

11. See chapter 3.

12. Gehlbach, F. R. 2008. "Eastern Screech Owl: Life History, Ecology, and Behavior in the Suburbs and Countryside." 2nd ed. *The W. L. Moody Jr. Life History Series*, no. 16. College Station: Texas A&M University Press.

13. See chapter 4.

14. See chapter 14.

15. Sumasgutner, P., E. Nemeth, G. Tebb, H. W. Krenn, and A. Gamauf. 2014. "Hard Times in the City—Attractive Nest Sites but Insufficient Food Supply Lead to Low Reproduction Rates in a Bird of Prey." *Frontiers in Zoology* 11:48. http://www.frontiersinzoology.com/content/11/1/48.

16. Gahbauer, M. A., D. M. Bird, K. E. Clark, T. French, D. W. Brauning, and F. A. McMorris. 2015. "Productivity, Mortality, and Management of Urban Peregrine Falcons in Northeastern North America." *Journal of Wildlife Management* 79:10–19.

17. Berry, M. E., C. E. Bock, and S. L. Haire. 1998. "Abundance of Diurnal Raptors on Open Space Grasslands in an Urbanized Landscape." *Condor* 100:601–8.

18. See chapter 10.

19. See chapter 8.

20. Mannan, R. W., and C. W. Boal. 2000. "Home Range Characteristics of Male Cooper's Hawks in an Urban Environment." *Wilson Bulletin* 112:21–27.

21. Chiang, S. N., P. H. Bloom, A. M. Bartuszevige, and S. E. Thomas. 2012. "Home Range and Habitat Use of Cooper's Hawks in Urban and Natural Areas." In *Urban Bird Ecology and Conservation,* edited by C. A. Lepczyk and P. S. Warren. Studies in Avian Biology, no. 45. Berkeley: University of California Press. http://www.ucpress.edu/go/sab.

22. Boggie, M. A., and R. W. Mannan. 2014. "Examining Seasonal Patterns of Space Use to Gauge How an Accipiter Responds to Urbanization." *Landscape and Urban Planning* 124:34–42.

23. Roth, T. C., II, W. E. Vetter, and S. L. Lima. 2008. "Spatial Ecology of Wintering *Accipiter* Hawks: Home Range, Habitat Use, and the Influence of Bird Feeders." *Condor* 110:260–68.

24. Lövy, M., and J. Riegert. 2013. "Home Ranges and Land Use of Urban Long-Eared Owls." *Condor* 115:551–57.

25. Bennett, J. R., and P. H. Bloom. 2005. "Home Range and Habitat Use by Great Horned Owls (*Bubo virginianus*) in Southern California." *Journal of Raptor Research* 39:119–26.

26. Dykstra, C. R., J. L. Hays, F. B. Daniel, and M. M. Simon. 2001. "Home Range and Habitat Use of Suburban Red-Shouldered Hawks in Southwestern Ohio." *Wilson Bulletin* 113:308–16.

27. Rutz, C. 2006. "Home Range Size, Habitat Use, Activity Patterns and Hunting Behavior of Urban-Breeding Northern Goshawks *Accipiter gentilis.*" *Ardea* 94:185–202.

28. Sodhi, N. S., and L. W. Oliphant. 1992. "Hunting Ranges and Habitat Use and Selection of Urban-Breeding Merlins." *Condor* 94:743–49.

29. England, A. S., J. A. Estep, and W. R. Holt. 1995. "Nest-Site Selection and Reproductive Performance of Urban-Nesting Swainson's Hawks in the Central Valley of California." *Journal of Raptor Research* 29:179–86.

30. Harrold, E. S. 2003. "Barred Owl (*Strix varia*) Nesting Ecology in the Southern Piedmont of North Carolina." MS thesis, University of North Carolina–Charlotte.

31. Morrison, J. L., I. G. W. Gottlieb, and K. E. Pias. 2016. "Spatial Distribution and the Value of Green Spaces for Urban Red-Tailed Hawks." *Urban Ecosystems* 19:1373. doi:10.1007/s11252-016-0554-0.

32. Peery, M. Z. 2000. "Factors Affecting Interspecies Variation in Home-Range Size of Raptors." *Auk* 117:511–17.

33. Rolando, A. 2002. "On the Ecology of Home Range in Birds." *Revue d'Ecologie (La Terre et la Vie)* 57:53–73.

34. Skipper, B. R. 2013. "Urban Ecology of Mississippi Kites." PhD diss., Texas Tech University.

35. Boal, C. W., and R. W. Mannan. 1998. "Nest-Site Selection by Cooper's Hawks in an Urban Environment." *Journal of Wildlife Management* 62:864–71.

36. Dykstra, C. R., J. L. Hays, F. B. Daniel, and M. M. Simon. 2000. "Nest Site Selection and Productivity of Suburban Red-Shouldered Hawks in Southern Ohio." *Condor* 102:401–8.

37. Stout, W. E., R. K. Anderson, and J. M. Papp. 1998. "Urban, Suburban and Rural Red-Tailed Hawk Nesting Habitat and Populations in Southeast Wisconsin." *Journal of Raptor Research* 32:221–28.

38. Stout, W. E., S. A. Temple, and J. R. Cary. 2006. "Landscape Features of Red-Tailed Hawk Nesting Habitat in an Urban/Suburban Environment." *Journal of Raptor Research* 40:181–92.

39. Salvati, L., L. Ranazzi, and A. Manganaro. 2002. "Habitat Preferences, Breeding Success, and Diet of the Barn Owl (*Tyto alba*) in Rome: Urban versus Rural Territories." *Journal of Raptor Research* 36:224–28.

40. Coleman, J. L., D. M. Bird, and E. A. Jacobs. 2002. "Habitat Use and Productivity of Sharp-Shinned Hawks Nesting in an Urban Area." *Wilson Bulletin* 114:467–73.

41. White, C. M., N. J. Clum, T. J. Cade, and W. G. Hunt. 2002. "Peregrine Falcon (*Falco peregrinus*)." In *The Birds of North America*, edited by P. G. Rodewald. Ithaca: Cornell Lab of Ornithology. Accessed January 26, 2017. https://birdsna.org/Species-Account/bna/species/perfal.

42. Charter, M., I. Izhaki, A. Bouskila, Y. Leshem, and V. Penteriani. 2007. "Breeding Success of the Eurasian Kestrel (*Falco tinnunculus*) Nesting on Buildings in Israel." *Journal of Raptor Research* 41:139–43.

43. Rodríguez, A., J. J. Negro, J. Bustamante, and J. Antolín. 2013. "Establishing a Lesser Kestrel Colony in an Urban Environment for Research Purposes." *Journal of Raptor Research* 47:214–18.

44. Tella, J. L., F. Hiraldo, J. A. Donázar-Sancho, and J. J. Negro. 1996. "Costs and Benefits of Urban Nesting in the Lesser Kestrel." In *Raptors in Human Landscapes: Adaptations to Built and Cultivated Environments*, edited by D. Bird, D. Varland, and J. J. Negro, 53–60. San Diego: Academic Press.

45. Sodhi, N. S., P. C. James, I. G. Warkentin, and L. W. Oliphant. 1992. "Breeding Ecology of Urban Merlins (*Falco columbarius*)." *Canadian Journal of Zoology* 70:1477–83.

46. Ellis, D. H., T. Craig, E. Craig, S. Postupalsky, C. T. LaRue, R. W. Nelson, D. W. Anderson, C. J. Henny, J. Watson, B. A. Millsap, J. W. Dawson, K. L. Cole, E. M. Martin, A. Margalida, and P. Kung. 2009. "Unusual Raptor Nests around the World." *Journal of Raptor Research* 43:175–98.

47. Bierregaard, R. O., A. F. Poole, M. S. Martell, P. Pyle, and M. A. Patten. 2016. "Osprey (*Pandion haliaetus*)." In *The Birds of North America*, edited by P. G. Rodewald. Ithaca: Cornell Lab of Ornithology. Accessed January 26, 2017. https://birdsna.org/Species-Account/bna/species/osprey.

48. Bloom, P. H., and M. D. McCrary. 1996. "The Urban Buteo: Red-Shouldered Hawks in Southern California." In *Raptors in Human Landscapes: Adaptations to Built and Cultivated Environments*, edited by D. Bird, D. Varland, and J. J. Negro, 31–39. San Diego: Academic Press.

49. Rottenborn, S. C. 2000. "Nest-Site Selection and Reproductive Success of Urban Red-Shouldered Hawks in Central California." *Journal of Raptor Research* 34:18–25.

50. James, P. C. 1992. "Urban-Nesting of Swainson's Hawks in Saskatchewan." *Condor* 94:773–74.

51. Rutz, C., R. G. Bijlsma, M. Marquiss, and R. E. Kenward. 2006. "Population Limitation in the Northern Goshawk in Europe: A Review with Case Studies." *Studies in Avian Biology* 31:158–97.

52. Holt, J. B., Jr. 1996. "A Banding Study of Cincinnati Area Great Horned Owls." *Journal of Raptor Research* 30:194–97.

53. See chapter 11.

54. Shanahan, D. F., M. W. Strohbach, P. S. Warren, and R. A. Fuller. 2014. "The Challenges of Urban Living." In *Avian Urban Ecology: Behavioural and Physiological Adaptation*, edited by D. Gil and H. Brumm, 3–20. Oxford, UK: Oxford University Press.

55. Hindmarch, S., and J. E. Elliott. 2014. "Comparing the Diet of Great Horned Owls (*Bubo virginianus*) in Rural and Urban Areas of Southwestern British Columbia." *Canadian Field-Naturalist* 128:393–99.

56. Teta, P., C. Hercolini, and G. Cueto. 2012. "Variation in the Diet of Western Barn Owls (*Tyto alba*) along an Urban–Rural Gradient." *Wilson Journal of Ornithology* 124:589–96.

57. Cade, T. J., M. Martell, P. Redig, G. Septon, and H. Tordoff. 1996. "Peregrine Falcons in Urban North America." In *Raptors in Human Landscapes: Adaptations to Built and Cultivated Environments*, edited by D. Bird, D. Varland, and J. J. Negro, 3–13. San Diego: Academic Press.

58. Krone, O., R. Altenkamp, and N. Kenntner. "Prevalence of *Trichomonas gallinae* in Northern Goshawks from the Berlin Area of Northeastern Germany." *Journal of Wildlife Diseases* 41:304–9.

59. Rutz, C., M. J. Whittingham, and I. Newton. 2006. "Age-Dependent Diet Choice in an Avian Top Predator." *Proceedings of the Royal Society B* 273:579–86.

60. Zalewski, A. 1994. "Diet of Urban and Suburban Tawny Owls (*Strix aluco*) in the Breeding Season." *Journal of Raptor Research* 28:246–52.

61. Frimer, O. 1989. "Food and Predation in Suburban Sparrowhawks *Accipiter nisus* during the Breeding Season." *Dansk Ornithologisk Forenings Tidsskrift* 83:35–44.

62. de Baerdemaeker, A. 2004. "The City of Rotterdam as Breeding and Wintering Habitat for the Eurasian Sparrowhawk *Accipiter nisus*." *Takkeling* 12:223–36. [In Dutch with English summary].

63. Kübler, S., S. Kupko, and U. Zeller. 2005. "The Kestrel (Falco *tinnunculus* L.) in Berlin: Investigation of Breeding Biology and Feeding Ecology." *Journal of Ornithology* 146:271–78.

64. Estes, W. A., and R. W. Mannan. 2003. "Feeding Behavior of Cooper's Hawks at Urban and Rural Nests in Southeastern Arizona." *Condor* 105:107–16.

65. Boal, C. W., and R. W. Mannan. 1999. "Comparative Breeding Ecology of Cooper's Hawks in Urban and Exurban Areas of Southeastern Arizona." *Journal of Wildlife Management* 63:77–84.

66. Frimer, O. 1989. "Breeding Performance in a Danish Suburban Population of Sparrowhawks *Accipiter nisus.*" *Dansk Ornithologisk Forenings Tidsskrift* 83:151–56.

67. Blanco, G. 1994. "Seasonal Abundance of Black Kites Associated with the Rubbish Dump of Madrid, Spain." *Journal of Raptor Research* 28:242–45.

68. De Giacomo, U., and G. Guerrieri. 2008. "The Feeding Behavior of the Black Kite (*Milvus migrans*) in the Rubbish Dump of Rome." *Journal of Raptor Research* 42:110–18.

69. Kumar, N., D. Mohan, Y. Jhala, and V. Yadvendradev. 2014. "Density, Laying Date, Breeding Success and Diet of Black Kites *Milvus migrans govinda* in the City of Delhi (India)." *Bird Study* 61:1–8.

70. Riegert, J., M. Lövy, and D. Fainová. 2009. "Diet Composition of Common Kestrels *Falco tinnunculus* and Long-Eared Owls *Asio otus* Coexisting in an Urban Environment." *Ornis Fennica* 86:123–30.

71. Cavalli, M., A. V. Baladrón, J. P. Isacch, G. Martínez, and M. S. Bó. 2014. "Prey Selection and Food Habits of Breeding Burrowing Owls (*Athene cunicularia*) in Natural and Modified Habitats of Argentine Pampas." *Emu* 114:184–88.

72. Dykstra, C. R., J. L. Hays, M. M. Simon, and F. B. Daniel. 2003. "Behavior and Prey of Nesting Red-Shouldered Hawks in Southwestern Ohio." *Journal of Raptor Research* 37:177–87.

73. Figueroa R., R. A. and D. González-Acuña. 2006. "Prey of the Harris's Hawk (*Parabuteo unicinctus*) in a Suburban Area of Southern Chile." *Journal of Raptor Research* 40:164–68.

74. González-Acuña, D., E. Briones, K. Ardíles, G. Valenzuela-Dellarossa, S. Corales S., and R. A. Figueroa R. 2009. "Seasonal Variation in the Diet of the White-Tailed Kite (*Elanus leucurus*) in a Suburban Area of Southern Chile." *Journal of Raptor Research* 43:134–41.

75. Ditchkoff, S. S., S. T. Saalfeld, and C. J. Gibson. 2006. "Animal Behavior in Urban Ecosystems: Modifications Due to Human-Induced Stress." *Urban Ecosystems* 9:5–12.

76. Carrete, M., and J. L. Tella. 2011. "Inter-Individual Variability in Fear of Humans and Relative Brain Size of the Species Are Related to Contemporary Urban Invasion in Birds." *PLoS ONE* 6: e18859.

77. Rebolo-Ifrán, N., M. Carrete, A. Sanz-Aguilar, S. Rodríguez-Martínez, S. Cabezas, T. A. Marchant, G. R. Bortolotti, and J. L. Tella. 2015. "Links between Fear of Humans, Stress and Survival Support a Non-Random Distribution of Birds among Urban and Rural Habitats." *Scientific Reports* 5:13723. doi:10.1038/srep13723.

78. Cavalli, M., A. V. Baladrón, J. P. Isacch, L. M. Biondi, and M. S. Bó. 2016. "Differential Risk Perception of Urban and Rural Burrowing Owls Exposed to Humans and Dogs." *Behavioural Processes* 124:60–65.

79. Parker, J. W. 1999. "Raptor Attacks on People." *Journal of Raptor Research* 33:63–66.

80. Stout, W. E., R. N. Rosenfield, W. G. Holton, and J. Bielefeldt. 2006. "The Status of Breeding Cooper's Hawks in the Metropolitan Milwaukee Area." *Passenger Pigeon* 68:309–20.

81. Driscoll, T. G., and R. N. Rosenfield. 2015. "Polygyny Leads to Disproportionate Recruitment in Urban Cooper's Hawks." *Journal of Raptor Research* 49:344–46.

82. Hiraldo, F., J. J. Negro, and J. A. Donázar. 1991. "Aborted Polygyny in the Lesser Kestrel *Falco naumanni*." *Ethology* 89:253–57.

83. Rosenfield, R. N., S. A. Sonsthagen, W. E. Stout, and S. L. Talbot. 2015. "High Frequency of Extra-Pair Paternity in an Urban Population of Cooper's Hawks." *Journal of Field Ornithology* 86:144–52.

84. Negro, J. J., J. A. Donázar, and F. Hiraldo. 1992. "Copulatory Behavior in a Colony of Lesser Kestrels: Sperm Competition and Mixed Reproductive Strategies." *Animal Behaviour* 43:921–30.

85. Chipman, E. D., N. E. McIntyre, R. E. Strauss, M. C. Wallace, J. D. Ray, and C. W. Boal. 2008. "Effects of Human Land Use on Western Burrowing Owl Foraging and Activity Budgets." *Journal of Raptor Research* 42:87–98.

86. Bautista, L. M., J. T. García, R. G. Calmaestra, C. Palacín, C. A. Martín, M. B. Morales, R. Bonal, and J. Viñuela. 2004. "Effect of Weekend Road Traffic on the Use of Space by Raptors." *Conservation Biology* 18:726–32.

87. Donázar, J. A., A. Cortés-Avizanda, J. A. Fargallo, A. Margalida, M. Moleón, Z. Morales-Reyes, R. Moreno-Opo, J. M. Pérez-García, J. A. Sánchez-Zapata, I. Zuberogoitia, and D. Serrano. 2016. "Roles of Raptors in a Changing World: From Flagships to Providers of Key Ecosystem Services." *Ardeola* 63:181–234.

88. Plumpton, D. L., and D. E. Andersen. "Anthropogenic Effects on Winter Behavior of Ferruginous Hawks." *Journal of Wildlife Management* 62:340–46.

CHAPTER 3

Urban Raptor Communities: Why Some Raptors and Not Others Occupy Urban Environments

Clint W. Boal

WE LIVE IN A WORLD that is experiencing rapid landscape-level changes due to human activities. Indeed, the argument can be made that, from a wildlife perspective, the only form of habitat that is increasing is what we would call "urban habitat." Despite the negative implications of this, some wildlife species are attracted to, and may even flourish in, urban settings. This is due primarily to urbanization resulting in an altered landscape that, often unintentionally, provides resources that function as components of habitat.[1] This was largely overlooked by biologists until Steve Emlen's 1974 publication examining bird communities in and near Tucson, Arizona.[2] Subsequent to Emlen's work, interest in the phenomenon of urban wildlife grew, and countless articles have since documented ornithological research in urban areas.[3]

Urban wildlife occupies novel landscapes consisting of a mosaic of native and introduced plants and animals, human activities, and structures with differing accessibility to resources such as water and food. For example, Lubbock, Texas, is located in the southern Great Plains of North America. In the 1880s, the location was an arid, flat, featureless, shortgrass prairie. Today, with more than 200,000 residents, the city of Lubbock is functionally an "island" of woodland with plentiful urban lakes, all surrounded by a "sea" of cotton fields and grasslands. Where wolves once pursued bison and black-footed ferrets chased prairiedogs, there are now tree squirrels, blue jays, and green herons, species that never naturally occurred there. These are the types of changes that urbanization can bring not only to the landscape but also to wildlife communities.

How do these interesting and novel wildlife communities form? Regardless of location, wildlife species interact with each other. This may include inter- and intraspecific competition for food, mates, and nest sites and more lethal interactions in which one attempts to depredate, or avoid being preyed upon by, others. These interactions, at least in part, lead to the structure of the community.

At its most basic level, a "community" comprises three elements. First is the diversity of the species present. Second is the relative abundance of each of those species compared to the others in the community. Last are the ecological niches or, simply put, the role that each of those species takes within the community. For example, consider a hypothetical raptor community consisting of three species: Mississippi kites (*Ictinia mississippiensis*), red-tailed hawks (*Buteo jamaicensis*), and great horned owls (*Bubo virginianus*). Because all three are raptors, their ecological niche is that of a predator within the larger ecological community. However, this is further refined by how, what, and when they hunt. Mississippi kites are diurnally active aerial foragers and prey primarily on aerial insects. Red-tailed hawks are also diurnally active but primarily capture small- to medium-sized ground-dwelling mammals and reptiles. The great horned owl also primarily hunts small- to medium-sized mammals but is nocturnal. So at least at a coarse level, the niches of the kite and hawk are separated by both how they hunt (aerial versus terrestrial prey) and what they hunt (invertebrate versus vertebrate). Thus there would be little competition between the species, which could facilitate their coexistence. The niches of the hawk and owl are primarily separated by when they hunt (diurnal versus nocturnal), again possibly facilitating their coexistence.

From an ecological and conservation perspective, understanding these interactions is just as important in the novel setting of an urban landscape as in wild and remote areas. Several contemporary studies have explored general avian community structure in urban settings, but none have explicitly examined raptor communities.[3] Studies of urban raptors tend to focus on individual species.[4,5] However, inherent ecological niche separation and local abundances of resources (e.g., nest sites and prey species drawn to urban environments) allow

for combinations of species to co-occur in urban settings where their distributions overlap.[6,7] Because the distributions of individual species may encompass
wide latitudinal and longitudinal gradients, there is potential for co-occurrence
with different assortments of other raptors within differently structured biological communities. Reports focusing on co-occurring species in urban areas
frequently document only presence and differential habitat selection.[6,7] However,
an important aspect of community structure for raptors is food.[8,9,10] Although
some studies have examined the diet of individual raptor species in urban settings,[11,12,13] studies explicitly assessing how food habits may influence raptor
communities in urban areas have been quite limited.[14] This is understandable,
as few researchers have the logistical and financial resources required to investigate multiple raptor species, which even individually are challenging to study.

The paucity of quantitative information on the richness and abundance
of raptors in urban landscapes hampers any review of the topic. However, by
combining different sources of information, some assessments can be made. In
this chapter, I use expert opinion and synthesis of data from existing sources
to (1) examine the proclivity of different raptor species to occupy urban areas,
(2) examine possible correlations of urban occupancy with measures of raptor
body sizes, natural habitats used, and food habits, and (3) explore differences
in food habits of raptor species between urban and rural settings. To provide
focus, I restricted this examination to North American raptors. Additionally,
much of the assessment is focused on diurnal raptors due to greater availability
of data. When using the term *urban*, I do not consider the size of the urban area
as significant as the intensity of human presence; suburbs, high-density urban
areas, small towns, and big cities are all included. However, I explicitly exclude
agricultural areas, which have substantial anthropogenic impacts but low human
presence.

Species-Specific Patterns of Urban Use

Although many raptors use urban areas in some seasons, quantitative data documenting seasonality and frequency of occurrence for many species is lacking. To
explore raptor distribution and tendency toward using or avoiding urban areas, I
compiled data available from the eBird website (http://ebird.org/content/ebird/).
The eBird website is a citizen science program that allows participants to enter
bird observations into a publically accessible online database. The program was
initiated by the Cornell Lab of Ornithology and the National Audubon Society
in 2002 and has grown to include more than 100,000 participants.

For this assessment, I chose a sample of 14 state capitals with municipal populations of 100,000 or more, based on 2010 census data (table 3.1). Rather than a random selection, I chose capital cities that were well distributed longitudinally and latitudinally across the continental United States. I developed a list of diurnal raptor species whose winter and/or summer distributions overlapped each capital city.[15] I then queried the eBird database for observations of each possible raptor species in each city during January and February 2016 as evidence of presence during winter and in May and June 2016 as evidence of presence during the breeding season. Because the eBird program involves citizen science, there may be issues with occasional misidentification. To attempt to control for misidentifications, I only tallied a species as present within a season if it was reported by three different observers across the study period. From these data, I determined the species that were prevalent in urban areas and those that were not present. Additionally, I created an index of frequency of occurrence for each raptor species by calculating the percentage of cities where each was detected from among the cities its range overlapped.

Table 3.1. Sample state capitals, population, numbers of diurnal raptor species possibly present and the percentage of those species detected[a] in the cities during winter (January and February) and summer (May and June) 2016.

| City | Metropolitan population[b] | Winter | | Summer | |
		Species possible	Percent present	Species possible	Percent present
Phoenix, AZ	4,192,887	17	82.4	13	69.2
Little Rock, AR	877,091	12	50.0	9	77.8
Sacramento, CA	2,527,123	15	86.7	11	72.7
Denver, CO	2,552,195	13	53.8	15	46.7
Hartford, CT	1,212,381	12	41.7	9	44.4
Tallahassee, FL	367,413	12	75.0	11	72.7
Boise, ID	616,561	12	75.0	14	64.3
Lincoln, NE	302,157	12	33.3	10	50.0
Columbus, OH	1,967,066	12	75.0	11	63.6
Oklahoma City, OK	1,252,987	13	61.5	12	58.3
Salem, OR	390,738	13	53.8	10	40.0
Columbia, SC	913,797	12	58.3	11	27.3
Austin, TX	1,716,291	16	75.0	14	78.6
Madison, WI	561,505	12	50.0	12	58.3

a. Calculated from eBird data.
b. 2010 population census.

Not all diurnal North American raptor species overlapped with the sample cities. To acquire use estimates for those that did not, I solicited the opinions of knowledgeable raptor researchers on the seasonality and proclivity of urban use by all North American diurnal raptors. Including myself, I collected the assessments of 13 experts with a geographical representation across the United States of the Pacific Northwest and Pacific Southwest, the Desert Southwest, central Rockies, the Gulf Coast, upper Midwest, and the Northeast. Furthermore, several experts had experience and knowledge from multiple geographic regions.

I categorized season of use as year-round, breeding, wintering, and migration. If a species was reported in both breeding and migration or wintering and migration, it was coded as breeding or wintering, respectively. The estimated proclivity of use was, of necessity, qualitative. Species deemed as absent from urban areas or occurring in them incidentally were categorized as 1 (never or incidental). Those species believed to only use urban areas on rare occasions were categorized as 2 (rare). Species known to use urban areas frequently enough that researchers would not be surprised to find them in such landscapes were categorized as 3 (frequent). Those species that occur in urban areas so frequently that researchers would be more surprised to not find them were considered as 4 (common). Degree of use and seasonality of use were expected to vary across any given species' distribution; for example, Mississippi kites are common nesting birds in cities of the southern Great Plains but much less frequent in cities of the southeastern United States. Seasonally, the species is present in urban areas during migration and the breeding season but absent during winter due to its migration to South America.

I conducted a correlation analysis between the frequency of occurrence index derived from the eBird data set and the estimated-use index developed from the expert solicitation. This was to assess how well the expert solicitation represented observed occurrences of raptors in urban areas. I found a strong positive correlation ($r^2 = 0.7387$, $P < 0.0001$) between the eBird data and expert opinion. This suggests that the expert opinion was reliable and justified applying their estimates of urban use to analyses so as to include raptor species that did not overlap with sample cities.

Among 31 diurnal raptor species, 17 species had distributional ranges that overlapped at least 50 percent of the sample cities in one or both seasons; 10 overlapped all sample cities (table 3.2). Turkey vultures (*Cathartes aura*), black vultures (*Coragyps atratus*), Cooper's hawks (*Accipiter cooperii*), sharp-shinned hawks (*Accipiter striatus*), red-tailed hawks, Swainson's hawks (*Buteo swainsoni*), bald eagles (*Haliaeetus leucocephalus*), American kestrels (*Falco sparverius*), and

Table 3.2. Detections of 31 diurnal raptor species among 14 sample cities during winter (January and February) and summer (May and June) 2016, as derived from the eBird database.

Species	Winter		Summer	
	Cities possible[a]	Presence detected[a]	Cities possible[a]	Presence detected[a]
Turkey vulture	8	6	14	14
Black vulture	8	7	5	3
Northern harrier	14	3	9	1
Osprey	8	4	14	9
White-tailed kite	4	1	2	1
Mississippi kite	np	np	5	5
Swallow-tailed kite	np	np	2	1
Sharp-shinned hawk	14	13	5	1
Cooper's hawk	14	14	14	11
Northern goshawk	5	0	4	0
Common black hawk	np	np	1	0
Harris's hawk	1	1	1	1
Zone-tailed hawk	1	1	2	0
Gray hawk	np	np	1	0
Red-shouldered hawk	10	6	10	6
Short-tailed hawk	nd	nd	nd	nd
Broad-winged hawk	np	np	10	5
Swainson's hawk	np	np	7	6
White-tailed hawk	nd	nd	nd	nd
Red-tailed hawk	14	14	14	13
Ferruginous hawk	7	0	2	0
Rough-legged hawk	10	0	np	np
Golden eagle	5	1	5	0
Bald eagle	14	12	13	7
Peregrine falcon	14	7	11	4
Gyrfalcon	nd	nd	nd	nd
Prairie falcon	8	6	3	0
American kestrel	14	11	14	7
Merlin	14	9	3	0
Aplomado falcon	nd	nd	nd	nd
Crested caracara	1	1	1	1

a. np = not present due to seasonal migration away from the area; nd = no data available due to species distribution not overlapping with 14 sample cities.

prairie falcons (*Falco mexicanus*) all occurred in at least 75 percent of the sample cities within their individual distributions during the summer, winter, or both. However, some species (e.g., sharp-shinned hawk, prairie falcon) occupied cities more during winter than during the breeding season. In contrast, ferruginous hawks (*Buteo regalis*), rough-legged hawks (*Buteo lagopus*), and northern harriers (*Circus hudsonius*), all open-country raptors, were either very rare or not detected in the sample cities their ranges overlapped for winter and breeding seasons. The remaining 14 species overlapped with few cities but were either present in all (e.g., Mississippi kite, Harris's hawk [*Parabuteo unicinctus*]) or almost completely absent (e.g., northern goshawk [*Accipiter gentilis*], zone-tailed hawk [*Buteo albonotatus*]; table 3.2).

Data were unavailable for swallow-tailed kites (*Elanoides forficatus*), white-tailed hawks (*Geranoaetus albicaudatus*), short-tailed hawks (*Buteo brachyurus*), aplomado falcons (*Falco femoralis*), and gyrfalcons (*Falco rusticolus*), as none of the sample cities were within the normal distribution for the first four, and the occurrence of gyrfalcons is so infrequent across its winter range as to be inconsequential. Experts opined that white-tailed hawks, aplomado falcons, and gyrfalcons are all open-country raptors that rarely, if ever, use urban areas. The swallow-tailed kite and short-tailed hawk, in contrast, are woodland raptors, both of which were considered by experts as being rare users of urban areas.

Overall, it appears that urban areas are used primarily by woodland and some mixed-habitat species or by those that key in on special resources. For example, many cities are built along watercourses, which are favored by bald eagles and ospreys (*Pandion haliaetus*) as foraging habitat. Open-country raptors in general, however, appear to avoid urban settings.

Relationships of Food Habits, Raptor Size, and Natural Habitat to Urban Use

A number of variables may influence whether a given raptor species occupies or avoids urban areas. To investigate some of these, I examined the relationships between the estimate of urban use (EUU) and a suite of variables associated with food habits. Data on food habits were obtained for each raptor species from published resources.[9,10,17] For species not included in these reports, I used the same methods to compile a suite of diet variables from other sources.[18,19,20,21,22,23,24,25,26] For methodological consistency, I applied the same criteria used by others for including a study: a minimum of 50 identified prey individuals, with all vertebrates identified to genus and invertebrates identified to order.[9] Additionally, I excluded obligate carrion feeders from the analysis. I then examined correlations between EUU and (1) the number of prey species, (2) the number of prey classes, (3) the ratio of the average mass of prey used by a raptor species to the average mass of that raptor, and (4) the average mass for both sexes, as an index to the size of each raptor species. I also examined urban occupancy with respect to (5) the geometric mean weight of prey, to account for skewed distributions in size of different prey used by any given raptor species;[9,10] (6) food niche breadths (FNBs), calculated using the Levin's index;[16] and (7) a standardized FNB of prey species, to account for geographic variation in prey species.[9]

In addition to prey selection, a raptor may use or avoid urban areas based on an innate template for "habitat" recognition and how plastic that template may be. To examine the role that habitat recognition may have on urban use, I conducted a correlation analysis between "natural" habitat use and EUU. Natural habitat used can differ between the breeding season, winter, and migration. Further, it can be broad or narrow depending on the species. To simplify the analysis, I used the very coarse habitat descriptors provided for each species by the Cornell Lab of Ornithology (https://www.allaboutbirds.org/). I pooled similar categories such as "scrub" and "desert." I then ranked each habitat type progressively from most open to least open in context of woody vertical structure: 1 = open (e.g., grasslands, marshlands), 2 = mixed (e.g., savannah, deserts), 3 = riparian woodlands, 4 = open woodlands, and 5 = forest. Data were missing from the website for three species so, based on personal knowledge, I categorized general habitat for short-tailed hawks as 4, common black-hawks (*Buteogallus anthracinus*) as 3, and aplomado falcons as 1.

Six of the eight variables I examined provided little value as possible predictors of urban use by diurnal raptors (table 3.3). However, the EUU was positively correlated to an increasing breadth of species in the diet (standardized FNB $P = 0.019$). This appears to be a factor only at the species level, as the correlation with the coarser level of prey classes (e.g., mammal or bird) was not significant ($P = 0.933$). In addition, the extent of urban use was also highly correlated ($P = 0.011$) to the type of natural habitat normally used (table 3.3). Specifically, those with frequent to common EUU tended to be those species that normally inhabit open woodlands and forested lands. In contrast, those species with incidental to rare EUU tended to normally occupy open country such as prairies, shrub-steppes, and deserts. This was consistent with the previous section's examination of distribution mapping and occurrence in urban areas.

Table 3.3. Correlation analysis of estimated level of urban use by raptors[a] with derived predictor variables.

Variable	*n*	*R*	*P*
Raptor mass	29	0.00119	0.859
Prey-predator mass ratio	23	0.10044	0.141
Geometric mean prey weight	23	0.02815	0.444
Number of prey classes	26	0.00123	0.865
Number of prey species	23	0.00010	0.964
Food niche breadth—class	26	0.00029	0.933
Food niche breadth—standardized	23	0.23452	0.019
Native habitat[b]	27	0.23309	0.011

a. Never/incidental, rare, frequent, common as estimated by expert elicitation.
b. Open, brushland, riparian woodland, open woodland, forest.

Comparison of Food Habits in Urban and Rural Settings

As noted in chapter 2, urban areas may hold different suites of prey species than natural areas, suggesting an urban-dwelling raptor may use different prey than conspecifics in natural areas. Indeed, the extent to which a raptor species is flexible in prey use may predispose them to occupy or avoid urban areas. This question has received surprisingly little quantitative assessment, with few studies comparing diet between urban and rural raptors during the same period. I compiled prey-use data from the studies that report food habits for raptors in urban and rural areas during the same time period and calculated species-level FNBs in urban and rural settings and diet overlap (%) between the two for each raptor species using Morisita's method.[11,12,13,27,28,29,30,31] For studies looking at gradients from urban to rural, I only used data for locations that could be clearly differentiated as urban and suburban or rural.

The percentage of dietary overlap between urban and rural conspecifics ranged from as low as 0 percent for barn owls (*Tyto alba*) to as high as 34.6 percent for great horned owls (table 3.4). FNB tended to be broader in urban areas for larger raptors but narrower for the smaller Eurasian kestrel (*Falco tinnunculus*) and the burrowing owl (*Athene cunicularia*; table 3.4). This may be due, at least in part, to the invertebrate-dominated diets of these smaller raptors and the coarseness of invertebrate identification, which was to the level of order only. Additionally, in some cases, there is a shift from native to nonnative prey in urban areas. Although voles (*Microtus* spp.) dominated barn owl and great horned owl diets in British Columbia, the proportion of introduced rodents

Table 3.4. Food niche breadth (FNB) at the species level and overlap of diet data, proportions, and percentage difference of avian prey in the diet from data collected at the same time periods for raptors in rural and urban locations.

Species[a]	Rural FNB[b]	Urban FNB[b]	Diet overlap (%)[c]	Birds in diet (%)		Difference (%)
				Rural	Urban	
Cooper's hawk	4.68	5.15	7.2	71.4	100.0	33
Eurasian kestrel	6.89	3.76	10.5	0.0	1.7	200
Eurasian kestrel	6.30	2.64	8.0	32.0	72.3	77
Tawny owl	6.09	6.32	4.5	9.1	59.7	147
Burrowing owl	4.78	3.6	20.2	0.7	2.9	122
Great horned owl	1.45	3.58	34.6	4.1	10.6	88
Barn owl	1.81	3.39	0.0	0.4	31.0	195
Barred owl[d]	na	na	na	5.4	58.3	166
Tawny owl[d]	na	na	na	2.4	12.1	134

a. Data from Campbell et al. 1987, Zalewski 1994, Tella et al. 1996, Estes and Mannan 2003, Kubler et al. 2005, Cauble 2008, Mrykalo et al. 2009, Grzedzicka et al. 2013; data analyzed only for prey from clear rural or urban settings.
b. Pianka's food niche breadth.
c. Morisita's index of diet overlap.
d. Species level data not available for calculation of food niche breadth.

(*Rattus* spp., *Mus musculus*) in their diets increased significantly in relation to increasing urbanization.[32,33] Urban-dwelling merlins (*Falco columbarius*) in Saskatoon, Saskatchewan, preyed primarily on the introduced house sparrow (*Passer domesticus*) and captured native horned larks (*Eremophila alpestris*) more frequently than would be expected based on their relative abundance.[34] The increases in abundance of a prey species resulted in increased selection of that species by merlins; in their case, this was primarily the house sparrow, which is substantially less abundant away from anthropogenic features.[34] Similarly, urban Cooper's hawks in Tucson, Arizona, captured more birds than rural hawks, although the species taken were primarily native dove species, which were more abundant in the urban area.[13] Additionally, urban-dwelling Mississippi kites shifted their diet to more abundant birds when the abundance of their normal invertebrate prey was reduced due to drought.[35]

Interestingly, in all six reviewed cases, there was a substantial increase in the proportion of birds in the diet of the urban raptors compared to their rural counterparts (table 3.4). Although the percentage of difference is high for the Eurasian kestrel and burrowing owl, the actual amount of avian prey in these species' diets is very small. However, the increases in the proportion of avian prey observed among the large urban raptors were substantial. For example, there were 147 percent, 166 percent, and 195 percent increases in the proportion of birds in the diet of urban tawny owls (*Strix aluco*), barred owls (*Strix varia*), and barn owls, respectively, compared to those in rural areas (table 3.4). Urban areas are generally characterized as having increased densities of avian species compared to natural areas.[3] Whether avian prey attracts some raptors to occupy urban areas or other features attract raptors that then take advantage of abundant avian prey remains unknown and could very well be specific to the individual raptor species. Regardless, there is an apparent pattern of urban-dwelling raptors shifting to include more avian prey in their diets compared to conspecifics in rural areas.

Conclusions

The limited availability of data and the use of solicited expert opinion necessitated that my analysis be rather coarse. Given that caveat, I found two factors that apparently increase the likelihood of a given raptor species using urban areas. First is their breadth of prey selection. This makes intuitive sense; raptors that have narrow prey selection may not find suitable prey or will be poorly suited to use the alternate prey that are available in urban settings. For example, ferruginous hawks specialize on rabbits and large, primarily burrowing rodents. Urban areas generally do not favor the presence of these animals in numbers sufficient to facilitate occupancy by ferruginous hawks. In contrast, a raptor that preys on a variety of species, such as a Cooper's hawk, would likely find an abundance of suitable prey in an urban setting, even if they are introduced species. In addition, many raptor species occupying urban areas take advantage of the increased number of birds present, even if birds are not normally a large component of their diet.

Second, the species' typical (natural) habitats apparently also influence whether they are likely to use urban areas. The features of a given urban area may be generally consistent with, or differ dramatically from, those in adjacent rural areas.[36,37] Along with this, the availability of resources, such as nesting areas and prey abundance, in urban areas may be attractive to some raptor species and not others. It is intuitively logical that a woodland species that forages primarily on birds, such as the Cooper's hawk, may inhabit an urban landscape rich in trees, provided sufficient prey is available (whether typical prey or alternative species to which it can shift). In contrast, it is also intuitive that an open-country raptor, such as a ferruginous hawk, would likely not be inclined to inhabit the same urban area.

Another important factor that contributes to whether a species can be urban, and one I was unable to address in my correlation analysis due to limited data, is the species' sensitivity to anthropogenic activities.[38,39] For example, Harris's hawks generally appear tolerant of human activities,[40] whereas white-tailed hawks are so sensitive to disturbance that they will flee their nest when human activity occurs as far as 500 m away.[41] Further, a particularly interesting case is the rarity with which the forest-dwelling northern goshawks use urban areas in North America (table 3.1). This is perplexing given that the closely related Cooper's hawks and sharp-shinned hawks readily take to urban areas. Even more confounding is that northern goshawks in Europe and Japan are not uncommon in urban areas. The northern goshawk will be an interesting species to monitor into the future to see if it too will expand its activities into urban areas in North America.

There is an abundance of intriguing research questions to be addressed concerning raptor use of, and activities in, urban areas. For North American raptors occupying urban areas, the majority of research has focused on nesting habitat and productivity; the in-depth ecological studies of Cooper's hawks and red-shouldered hawks (*Buteo lineatus*) in different cities and latitudes are notable exceptions.[42,43] Fortunately, similar ecological research is being undertaken for other species, such as barred owls and Mississippi kites.[44,45] Currently, we have a generally good, if qualitative, idea of which diurnal species will regularly use urban areas; unfortunately, little has been reported for the majority of nocturnal species. As research on urban raptors progresses, we will no doubt develop a better understanding of why some raptors but not others will use urban areas and how different combinations of resources and inter- and intraspecific interactions influence community structure. Perhaps more important will be developing an understanding of how urban areas may be used for the conservation of at least some birds of prey, as has already occurred for peregrine falcons.[46]

Acknowledgments

I thank S. DeStefano, C. Dykstra, A. Fish, and K. Watson for providing thoughtful reviews of this manuscript. Expert opinions regarding occurrence of raptors in urban landscapes were kindly provided by S. Ausubel, P. Bloom, W. Clark, J. Coulson, E. Deal, J. Dwyer, A. Fish, R. Mannan, M. Martell, E. Mojica, M. Normandia, and M. Tincher. Any use of trade, firm, or product names is for descriptive purposes only and does not imply endorsement by the US government.

Literature Cited

1. Hall, L. S., P. R. Krausman, and M. L. Morrison. 1997. "The Habitat Concept and a Plea for Standard Terminology." *Wildlife Society Bulletin* 25:173–82.

2. Emlen, J. T. 1974. "An Urban Bird Community in Tucson, Arizona: Derivation, Structure, Regulation." *Condor* 76:184–97.

3. Marzluff, J. M. 2016. "A Decadal Review of Urban Ornithology and a Prospectus for the Future." *Ibis* 159:1–13.

4. Bird, D. M., D. E. Varland, and J. J. Negro. 1996. *Raptors in Human Landscapes: Adaptations to Built and Cultivated Environments*. San Diego: Academic Press.

5. Mannan, R. W., and C. W. Boal. 2004. "Birds of Prey in Urban Landscapes." In *People and Predators*, edited by N. Fascione, A. Delach, and M. E. Smith, 105–17. Washington, DC: Island Press.

6. Mannan, R. W., C. W. Boal, W. J. Burroughs, J. W. Dawson, T. S. Estabrook, and W. S. Richardson. 2000. "Nest Sites of Five Raptor Species along an Urban Gradient." In *Raptors at Risk: Proceedings of the V World Conference on Birds of Prey and Owls*, edited by R. D. Chancellor and B.-U. Meyburg, 447–53. Berlin, Germany: World Working Group on Birds of Prey and Owls; Surrey, BC: Hancock House Publishers.

7. Hogg, J. R., and C. H. Nilon. 2015. "Habitat Associations of Birds of Prey in Urban Business Parks." *Urban Ecosystems* 18:267–84.

8. Jaksić, F. M., and H. E. Braker. 1983. "Food-Niche Relationships and Guild Structure of Diurnal Birds of Prey: Competition versus Opportunism." *Canadian Journal of Zoology* 61:2230–41.

9. Marti, C. D, E. Korpimäki, and F. M. Jaksić. 1993a. "Trophic Structure of Raptor Communities: A Three-Continent Comparison and Synthesis." *Current Ornithology* 10:47–137.

10. Marti, C. D., K. Steenhof, M. N. Kochert, and J. S. Marks. 1993b. "Community Trophic Structure: The Roles of Diet, Body Size, and Activity Time in Vertebrate Predators." *Oikos* 67:6–18.

11. Zalewski, A. 1994. "Diet of Urban and Suburban Tawny Owls (*Strix aluco*) in the Breeding Season." *Journal of Raptor Research* 28:246–52.

12. Tella, J. L., F. Hiraldo, J. A. Donazar-Sancho, and J. J. Gegro. 1996. "Costs and Benefits of Urban Nesting in Lesser Kestrel." In *Raptors in Human Landscapes: Adaptations to Built and Cultivated Environments*, edited by D. M. Bird, D. E. Varland, and J. J. Negro, 53–60. San Diego: Academic Press.

13. Estes, W. A., and R. W. Mannan. 2003. "Feeding Behavior of Cooper's Hawks at Urban and Rural Nests in Southeastern Arizona." *Condor* 105:107–16.

14. Riegert, J., M. Lövy, and D. Fainová. 2009. "Diet Composition of Common Kestrels *Falco tinnunculus* and Long-Eared Owls *Asio otus* Co-Existing in an Urban Environment." *Ornis Fennica* 86:123–30.

15. Sibley, D. A. 2014. *The Sibley Guide to Birds*. 2nd ed. New York: Alfred A. Knopf.

16. Krebs, C. J. 1999. *Ecological Methodology*. 2nd ed. Menlo Park: Addison Wesley Longman.

17. Snyder, N. F. R., and J. W. Wiley. 1976. "Sexual Size Dimorphism in Hawks and Owls of North America." *Ornithological Monographs* 20:1–95.

18. Glinski, R. L., and R. D. Ohmart. 1983. "Breeding Ecology of the Mississippi Kite in Arizona." *Condor* 85:200–207.

19. Millsap, B. A. 1981. "Distributional Status of Falconiformes in Westcentral Arizona: With Notes on Ecology, Reproductive Success and Management." US Bureau of Land Management, Technical Note No. 355. Phoenix District Office, Phoenix, AZ.

20. Farquhar, C. C. 2009. "White-Tailed Hawk (*Geranoaetus albicaudatus*)." In *The Birds of North America*, edited by P. G. Rodewald. Ithaca: Cornell Lab of Ornithology. Accessed January 26, 2017. https://birdsna.org/Species-Account/bna/species/comblh1.

21. Schnell, J. H. 1994. "Common Black Hawk (*Buteogallus anthracinus*)." In *The Birds of North America*, edited by P. G. Rodewald. Ithaca: Cornell Lab of Ornithology. Accessed January 26, 2017. https://birdsna.org/Species-Account/bna/species/comblh1.

22. Kennedy, P. L., D. E. Crowe, and T. F. Dean. 1995. "Breeding Biology of the Zone-Tailed Hawk at the Limit of Its Distribution." *Journal of Raptor Research* 29:110–16.

23. Meyer, K. D., and M. W. Collopy. 1995. "Status, Distribution, and Habitat Requirements of the American Swallow-Tailed Kite (*Elanoides forficatus*) in Florida." Nongame Wildlife Program Project GFC-87-025. Florida Game and Fresh Water Fish Commission, Tallahassee, FL.

24. Howell, D. L., and B. R. Chapman. 1998. "Prey Brought to Red-Shouldered Hawk Nests in the Georgia Piedmont." *Journal of Raptor Research* 32:257–60.

25. Johnson, R. R., R. L. Glinski, and S. W. Matteson. 2000. "Zone-Tailed Hawk (*Buteo albonotatus*)." In *The Birds of North America*, edited by P. G. Rodewald. Ithaca: Cornell Lab of Ornithology. Accessed January 26, 2017. https://birdsna.org/Species-Account/bna/species/comblh1.

26. Strobel, B. N. 2007. "Reproductive Success, Habitat Selection, and Nestling Diet of the Texas Red-Shouldered Hawk in South Texas." MS thesis, Texas Tech University.

27. Campbell, R. W., D. A. Manuwal, and A. S. Harestad. 1987. "Food Habits of the Common Barn-Owl in British Columbia." *Canadian Journal of Zoology* 65:578–86.

28. Kübler, S., S. Kupko, and U. Zeller. 2005. "The Kestrel (Falco *tinnunculus* L.) in Berlin: Investigation of Breeding Biology and Feeding Ecology." *Journal of Ornithology* 146:271–78.

29. Cauble, L. C. 2008. "The Diets of Rural and Suburban Barred Owls (*Strix varia*) in Mecklenburg County, North Carolina." MS thesis, University of North Carolina–Charlotte.

30. Mrykalo, R. J., M. M. Grigione, and R. J. Sarno. 2009. "A Comparison of Available Prey and Diet of Florida Burrowing Owls in Urban and Rural Environments: A First Study." *Condor* 111:556–59.

31. Grzedzicka, E., K. Kus, and J. Nabielec. 2013. "The Effect of Urbanization on the Diet Composition of the Tawny Owl (*Strix aluco* L.)." *Polish Journal of Ecology* 61:391–400.

32. Hindmarch, S., and J. E. Elliott. 2015. "A Specialist in the City: The Diet of Barn Owls Along a Rural to Urban Gradient." *Urban Ecosystems* 18:477–88.

33. Hindmarch, S., and J. E. Elliott. 2014. "Comparing the Diet of Great Horned Owls (*Bubo virginianus*) in Rural and Urban Areas of Southwestern British Columbia." *Canadian Field-Naturalist* 128:393–99.

34. Sodhi, N. S., and L. W. Oliphant. 1993. "Prey Selection by Urban-Breeding Merlins." *Auk* 110:727–35.

35. Welch, B. C., and C. W. Boal. 2015. "Prey Use and Provisioning Rates of Urban-Nesting Mississippi Kites in West Texas." *Journal of Raptor Research* 49:141–51.

36. See chapter 2.

37. See chapter 4.

38. Steidl, R. J., and B. F. Powell. 2006. "Assessing the Effects of Human Activities on Wildlife." *George Wright Forum* 23:50–58.

39. Carrete, M., and J. L. Tella. 2011. "Inter-Individual Variability in Fear of Humans and Relative Brain Size of the Species Are Related to Contemporary Urban Invasion in Birds." *PLoS ONE* 6: e18859.

40. See chapter 9.

41. Haralson, C. L. 2008. "Breeding Ecology, Nest Site Selection, and Human Influence of White-Tailed Hawks on the Texas Barrier Islands." MS thesis, Texas Tech University.

42. See chapter 7.

43. See chapter 8.

44. See chapter 10.

45. See chapter 6.

46. See chapter 13.

CHAPTER 4

Demography of Raptor Populations in Urban Environments

R. William Mannan and Robert J. Steidl

L ANDS MODIFIED BY HUMANS VARY widely in their environmental features, from areas that retain much of their natural character to areas in the urban core that retain little natural character. For example, a city that was built on lands that were once covered by a forest might be dominated at its core by pavement and buildings and support only a few trees, many of which are nonnative species. Areas surrounding the urban core usually are residential neighborhoods composed of private homes and small parks, where trees are more common. On the outskirts of the city are exurban areas dominated by natural forest vegetation with only scattered houses and other human structures. Consequently, urban areas often represent a gradient of development[1] that spans an array of natural and anthropogenic features, many of which influence the probability of an area being inhabited by a species (figure 4.1). Some raptors are capable of inhabiting urban environments at one or more points along this gradient, provided the areas support their specific habitat resources and match their tolerance of human activity.[2,3]

Raptors can establish populations in urban areas naturally (e.g., northern goshawks [*Accipiter gentilis*] in Germany),[4] when facilitated through reintroduction

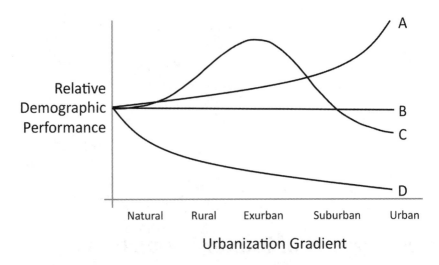

Figure 4.1. A conceptual model of potential demographic responses of raptors to urbanization. Line A illustrates the demographic performance of species that find high-quality habitat in urban areas; line B illustrates the demographic performance of species whose habitat quality is similar in urban and nonurban areas; line C illustrates the response of species whose demographic performance varies widely along the urban gradient; and line D illustrates the response of species whose demographic performance declines in urban areas.

programs (e.g., hacking of peregrine falcons [*Falco peregrinus*]),[5] or by persisting in areas that are overtaken by urban or suburban sprawl. Once established in urban areas, as discussed in chapters 2 and 3, these populations sometimes benefit from reduced predator densities, fewer competitors, reduced rates of persecution by humans, and access to abundant and stable sources of food, nest sites, and other key resources.[2,6] In contrast, however, urban environments often expose raptors to novel processes and structures that increase mortality rates, including electrocution, poisoning, exotic diseases, and collisions with anthropogenic structures such as windows, automobiles, and power lines.[7] The interplay between access to abundant resources and exposure to novel features and processes can produce population dynamics that are complex, location-specific, and markedly different from those expressed in natural environments.

In this chapter, we describe demography of raptor populations that inhabit areas along the gradient of human-modified environments, including urban, suburban, exurban, and rural areas, which we contrast with demography of raptor populations inhabiting natural areas. Such comparisons will be constrained, however, because the amount and type of information available on demography of raptor populations in urban versus nonurban environments vary widely by species and region. For example, nest density and productivity (i.e., reproductive rate) of many raptors in urban and less-developed environments have been studied relatively well, but survival has been explored for only a few species, resulting in few studies that have described the effects of urbanization on the rate of population change (λ). Further, most studies have contrasted only two points along the urban gradient, which provides insight into the effects of urbanization on raptors along only narrow portions of this broad gradient. For species capable of inhabiting a wide range along the urban gradient, their demography could vary markedly outside of the particular areas studied, so inferences about differences in demography of raptors inhabiting urban and less-developed areas will be limited to the areas studied and their relative positions along the urban gradient (figure 4.1). For example, in and near New Mexico State University in Las Cruces, New Mexico, nests of burrowing owls (*Athene cunicularia*) in human-altered areas (i.e., on campus) were farther apart and produced more nestlings and fledglings than nearby nests in areas of natural desert vegetation.[8] In another study in the same region, burrowing owls in prairie dog colonies in natural grasslands nested at higher densities and had higher nest success than owls nesting in urban Las Cruces, but average fledging success was similar between the two areas.[9]

Ultimately, our intent is not to compile all available demographic information on raptors in urban environments but to illustrate the breadth and consequences of demographic responses of raptors to urbanization. To provide a framework for our ideas, we develop a conceptual model that describes the range of responses by raptors to urbanization across a range of environments; we limit our review to species that inhabit urban areas year-round or during the breeding season.

Raptors in Urban Environments

A species can inhabit areas where its resource needs are met, provided environmental conditions are within its tolerance limits for climatic extremes and human activities. More specifically, a species can complete a portion or all of its life history in areas where the abiotic environment is suitable and the biotic environment provides necessary resources. For many breeding raptors, key habitat resources often are relatively modest. Peregrine falcons, for example, require a nesting ledge that is often on a vertical surface positioned well above the ground, such as a sheer cliff. For species such as ospreys (*Pandion haliaetus*) and red-tailed hawks (*Buteo jamaicensis*), a key resource is a tree capable of supporting a nest that offers good visibility of the surrounding terrain. These species and others often are sufficiently flexible in their choice of nesting substrate that they will readily substitute structures that are common in urban environments for natural substrates, such as tall buildings and bridges for peregrine falcons and utility poles or electrical transmission towers for some tree-nesting raptors. When these key resources are introduced through the urbanization process, the probability of residency for these species increases when the urban environment also provides the other resources they need, especially prey (the "bedroom" and "pantry" referred to in chapter 1).

One researcher from Baylor University, Frederick Gehlbach,[6] developed a model to describe responses of raptors in general to urbanization, based on his observations of eastern screech-owls (*Megascops asio*) in urban and suburban areas of Texas. He described how urban areas can provide high-quality habitat for some raptors and predicted their demographic responses. In brief, he proposed that urban areas often provide environmental conditions that are more favorable and more stable than natural areas. Given that urban areas often have relatively moderate climates, abundant sources of permanent water, many potential nest sites, fewer predators and competitors, and abundant, stable prey populations, the demography of raptors in these areas should reflect those of animals inhabiting high-quality habitats. He suggested that raptors in urban areas should initiate nests earlier than in nonurban areas and have higher rates of adult survival and productivity, increased recruitment of fledglings, relatively stable dynamics, and higher nest densities. Additionally, higher temperatures in suburban areas, a result of the "heat-island effect," might function to increase productivity of raptor populations further due to an increase in availability of both ectothermic and endothermic prey, such as insects and birds, respectively.[10] This model accurately describes responses of some species of raptors to urban areas, but the variety and complexity of environments along the urban gradient, species-specific habitat requirements and tolerances to human disturbance, and the novel processes and structures in urban areas that can affect survival and recruitment suggest a wider array of potential demographic responses of raptors to urbanization. Therefore, we extend Gehlbach's model by expanding the range of possible responses of raptors along the urban gradient (figure 4.1). We begin by describing two species—Mississippi kites (*Ictinia mississippiensis*) and merlins (*Falco columbarius*), whose demographic responses match some of Gehlbach's predictions—and suggest that urban areas provide high-quality habitat for them (figure 4.1, line A). We then describe the demographic responses of other raptors to urbanization to illustrate the breadth of variation that exists.

Following a range extension west of the Mississippi River in the mid to late 1900s, Mississippi kites became the most abundant raptor inhabiting urban environments[11] and among the most successful demographically.[12,13] (Kites are covered in more detail in chapter 6.) Survival of breeding adults in urban areas is high,[12] and in both Kansas and Texas, productivity and nest success are higher in urban than in nonurban areas.[11,13] An abundant food supply, including birds, mammals, and insects, combined with numerous suitable nest trees and fewer predators, is likely the key reason for success of Mississippi kites in urban areas.[11]

Merlins began inhabiting urban areas in the late 1960s and early 1970s, and a 19-year study in Saskatoon, Saskatchewan, documented rapid growth and demography of one population.[14] Like Mississippi kites, merlins exhibited nesting density and productivity in this urban area that were among the highest reported for the species at that time. Factors that likely led to the success of this population included few natural predators and abundances of nest sites and prey.[14]

Demographic responses of other raptor species suggest that urban areas can provide habitat that is of similar quality to natural areas (figure 4.1, line B). Red-tailed hawks, for example, are common in urban areas throughout North America (e.g., Arizona,[3] Wisconsin,[15] New York[16]), and often their productivity in urban and nonurban areas is similar, at least for the few areas studied. Productivity of red-tailed hawks in urban areas of New York did not differ appreciably from hawks in rural areas of the midwestern and eastern United States, although nest density in rural areas tended to be higher.[16] In Wisconsin, productivity of red-tailed hawks did not differ among nests in urban, suburban, and rural areas.[15]

Similarly, productivity of red-shouldered hawks (*Buteo lineatus*) in urban and suburban areas of California in the mid to late 1980s[17] did not vary with the degree of urbanization around nests.[18] In addition, productivity of red-shouldered hawks in suburban areas of southwestern Ohio did not differ from productivity in rural areas.[19,20] Interestingly, structure of vegetation around nests in suburban and rural areas was also similar,[19,20] suggesting that hawks were able to find the nesting resources they require at different points along the urban gradient.

For raptors capable of inhabiting a broad span of the urban gradient, demographic responses sometimes vary along the gradient (figure 4.1, line C). Demography of burrowing owls, for example, varies with intensity of urban development. In Florida, nest density of burrowing owls increased until 45–60 percent of lots in a subdivision were developed.[21] Nest success was relatively uniform across the development gradient, but the number of young fledged per successful nest decreased when more than 60 percent of lots were developed.

Benefits accrued by burrowing owls from increased abundance of prey in highly developed areas were thought to be offset by human-caused nest failures.[21] Survival of adult and juvenile burrowing owls varied across the development gradient but were related inversely to each other; survival of juveniles was high in areas where survival of adults was low.[22]

Finally, for some raptor species, habitat quality can decline as the degree of urbanization increases (figure 4.1, line D), although the species may still inhabit urban areas. Barn owls (*Tyto alba*) are declining in abundance and their geographical range is decreasing in western Europe and North America as a result of increasing intensity of agricultural practices, including use of pesticides, conversion of grasslands (i.e., foraging habitat) to croplands, and reductions in the number of nest sites through removal of barns and old trees.[23] Other factors associated with the decline of barn owls increase with urbanization,[23] such as increased mortality from collisions with motor vehicles and loss and fragmentation of grasslands. In Fraser Valley, British Columbia, the amount of urban cover was the dominant factor associated with decreased fledging success in barn owls.[23] The relationship between increased urban cover and lower fledging success was unclear, but urbanization may reduce abundance or accessibility of some prey directly or indirectly[23] or may increase exposure to anticoagulant rodenticides when owls consume rodents from urban areas.[24] In contrast, fledging success of barn owls in urban and rural areas in Italy was similar, but occupied sites contained less urban cover than expected based on availability.[25]

Effects of Urbanization on Raptors Can Vary Regionally

The influence of urbanization on native plants, animals, and fundamental ecological processes likely will vary among geographic regions and with the type and magnitude of development. In the Sonoran Desert near Tucson, Arizona, for example, native woody vegetation is dominated by velvet mesquite (*Prosopis velutina*), palo verde (*Parkinsonia microphylla*), and creosote bush (*Larrea tridentata*). When land in this region is developed for residential housing, many of these native plants are replaced with nonnative species. These nonnatives differ markedly in structure from natives, including grass lawns, ornamental shrubs, and trees, the most common of which are *Eucalyptus* species, Aleppo pines (*Pinus halepensis*), and species of mesquite from the neotropics, all of which grow much taller than native trees. Furthermore, these landscapes typically are enriched perennially with fertilizer and water. As a consequence, the amount and diversity of vegetation in developed areas is much higher than in surrounding natural desert areas. In general, these changes increase the amount of nesting habitat for some raptors and the abundance of many potential prey species.

The type and magnitude of changes resulting from urbanization in desert systems are likely to be quite different from those that occur when development takes place in other regions. When deciduous forests of the eastern United States are developed, for example, structural diversity of vegetation could be reduced substantially, and rates of primary productivity may not increase after development, at least compared to desert systems. Thus regional variation in how urbanization influences vegetation and ecological processes—and therefore resources for raptors—may help explain why the same raptor species responds differently to urbanization in different regions.

EXAMPLE SPECIES: COOPER'S HAWK

The Cooper's hawk (*Accipiter cooperii*) is widespread in North America, inhabits many urban areas, and has been relatively well studied (and is covered in depth in chapter 7). Densities of nests of Cooper's hawks in urban areas are among the highest reported for the species,[26] and home ranges often are smaller than those reported in nonurban areas. For example, home-range sizes of Cooper's hawks in Tucson averaged 72.7 ha,[27] about 20 times smaller than those in relatively natural environments elsewhere.[28] The small home-range sizes for Cooper's hawks in urban environments are likely a product of abundant avian prey. Bird communities in urban areas generally are characterized by extraordinary densities and total biomass relative to more natural areas.[2,29] In short, Cooper's hawks in urban areas do not have to go far to find food.[27]

Productivity of Cooper's hawks in urban areas compared to nonurban areas varies by region, especially between the western and eastern United States. In Arizona, nest initiation is early (by 16 days) compared to nearby nonurban areas,[30] but in Wisconsin, nesting phenology does not differ between urban and nonurban areas.[26] Brood sizes in general tend to be relatively large for urban-nesting Cooper's hawks, but while they are larger in urban (compared to nearby nonurban) nests in Arizona and California, they are not in Wisconsin.[26]

Annual survival rate of breeding male Cooper's hawks is relatively high and was similar in urban (0.84) and nonurban areas (0.79) in Wisconsin[26] and similar to that reported for both breeding males and females in an urban-nesting population in Arizona (0.80).[31] Of interest is that survival of juvenile hawks in their first winter in an urban area in Arizona also was high (0.67 for males and females combined) compared to other predatory birds during dispersal.[31,32] The high rates of survival for both adult and juvenile birds likely is a product of the abundance of food (i.e., small birds) in urban areas.

EXAMPLE SPECIES: PEREGRINE FALCON

As the authors of chapter 13 discuss, peregrine falcon populations declined precipitously worldwide during the 1950s and 1960s in response to the adverse effects of organochlorine pesticides, specifically DDT and its metabolites.[33] Aggressive management to recover this species included introductions of captive-bred juveniles by fostering, cross-fostering, and hacking (i.e., releasing juveniles from hack boxes without adults).[33] Hack sites often were located in urban areas, which likely helped to foster establishment of peregrine falcon populations in many urban areas of North America. Population models based on a more than 20-year-mark recapture analysis of peregrine falcons in California indicated that urban populations, on average, were growing more rapidly (λ = 1.28) than rural populations (λ = 0.99), primarily due to higher rates of survival of first-year birds and overall productivity in urban areas.[33]

In contrast, an analysis of more than 800 nesting attempts by peregrine falcons from 1980 through 2006 in eastern North America (Ontario, Quebec, Massachusetts, New Jersey, and Pennsylvania) indicated that productivity did not differ appreciably between nests in urban and rural areas.[34] Differences in the effects of urbanization on peregrine falcons between the eastern United States and California might be explained by differences in the way urbanization influences different ecosystems. As noted above, urbanization likely enhances the overall productivity of arid areas (e.g., grasslands and deserts of California) more than mesic areas (e.g., hardwood forests in the eastern United States). As a result, peregrine falcons nesting in urban areas in the West may benefit from a higher abundance of food relative to nonurban areas, but that advantage might be absent in the eastern United States.

Novel Features in Urban Areas

Although urban areas confer many benefits to peregrine falcons and other raptors, they also pose threats from novel structures, such as plate glass windows and electrical transmission lines; novel organisms, including predators and diseases; and novel situations, such as moving vehicles, with which raptors did not evolve. These features can be significant causes of mortality[7] and potentially influence demography of raptors that inhabit urban environments, but their influence varies among species. For example, Harris's hawks (*Parabuteo unicinctus*) nest and hunt cooperatively in groups of up to seven individuals,[35] and in the early 2000s in Tucson, Arizona, an average of 1.87 hawks (usually adult females and fledglings) per nesting group were electrocuted annually.[36] In contrast, Cooper's hawks that nest in Tucson are less affected by electrocution (likely because they are smaller than Harris's hawks) but die regularly from collisions with windows and cars.[37]

In circumstances where mortality is exceptionally high and caused predominantly by anthropogenic factors, urban areas could function as ecological traps. An ecological trap is an area where the environmental cues used by animals to identify habitat are decoupled from the evolutionary forces that once linked those cues to habitat quality, therefore individuals are misled into settling in areas where productivity might be insufficient to offset mortality.[38,39] The powerful owl (*Ninox strenua*)—the largest owl in Australia, an apex predator, and thought to be a forest obligate—inhabits urban and suburban areas surrounding Melbourne.[40] As discussed in chapter 11, an analysis of resources thought to be important cues for triggering settlement (i.e., food and nest cavities) suggested that urban and suburban areas provided adequate prey (arboreal marsupials), but nest cavities were lacking.[40] Owls that resided in urban and urban-fringe areas, presumably attracted by the abundance of food, often did not breed, suggesting that urban areas were a potential ecological trap for this species.[40]

In the mid-1990s in Tucson, Arizona, trichomoniasis, an urban-related disease[41] caused by the protozoan *Trichomonas gallinae*, killed about 40 percent of the nestling Cooper's hawks produced each year but had little effect on hawks over 60 days of age.[37] Members of the family Columbidae are the primary hosts for *T. gallinae*, and nestlings acquired the disease from doves that their parents fed them. The high mortality of nestlings suggested that Tucson could function as an ecological trap for Cooper's hawks.[37] Subsequent research indicated that the population of Cooper's hawks in Tucson was growing and the city was thus not an ecological trap, primarily because the high rates of survival of hawks in their first year of life (if they survived the disease) offset the disease-related mortality among nestlings.[31] The pattern of survival of young Cooper's hawks in Tucson— low survival as nestlings and relatively high survival in their first year of life—is opposite that in more natural settings, where nestlings often survive to fledge but then die at relatively high rates during their first year of life, usually due to starvation. In Tucson, the same abundant source of food that resulted in nestling deaths was likely also the reason for relatively high survival of hawks in their first year of life.

Conclusions

Because urban areas vary widely in the ways that their environmental features differ from nearby natural areas, changes resulting from the urbanization process will favor some raptor species and disfavor others. In general, species are favored when they are less sensitive to the activities associated with development or human use of urban areas and when urbanization increases the availability of habitat resources.[42] Changes to environmental features and processes resulting from urbanization can vary geographically, however, which explains in part why some species inhabit urban areas in some geographic regions but not others and why the demographic responses to urbanization of the same species vary among regions. Therefore, predicting reliably whether an urban area might provide high-quality habitat for a particular raptor species will require understanding its specific resource requirements and sensitivity to human activities.

Although development reduces the amount of natural area, urban areas can sometimes provide habitat for species that are capable of tolerating the human footprint and exploiting resources that sometimes can be unusually abundant. Consequently, urban areas will play a role in conservation and even recovery of some species while contributing to concomitant reductions in others.

Literature Cited

1. Blair, R. B. 1996. "Land Use and Avian Species along an Urban Gradient." *Ecological Applications* 6:506–19.
2. Chace, J. F., and J. J. Walsh. 2006. "Urban Effects on Native Avifauna: A Review." *Landscape and Urban Planning* 74:46–69.
3. Mannan, R. W., C. W. Boal, W. J. Burroughs, J. W. Dawson, T. S. Estabrook, and W. S. Richardson. 2000. "Nest Sites of Five Raptor Species along an Urban Gradient." In *Raptors at Risk: Proceedings of the V World Conference on Birds of Prey and Owls*, edited by R. D. Chancellor and B.-U. Meyburg, 447–53. Berlin, Germany: World Working Group on Birds of Prey and Owls; Surrey, BC: Hancock House Publishers.
4. Rutz, C. 2008. "The Establishment of an Urban Bird Population." *Journal of Animal Ecology* 77:1008–19.
5. Caballero, I. C., J. M. Bates, M. Hennen, and M. V. Ashley. 2016. "Sex in the City: Breeding Behavior of Urban Peregrine Falcons in the Midwestern US." *PLoS ONE* 11(7): e0159054. doi:10.1371/journal.pone.0159054.
6. Gehlbach, F. R. 1996. "Eastern Screech Owls in Suburbia: A Model of Raptor Urbanization." In *Raptors in Human Landscapes: Adaptations to Built and Cultivated*

Environments, edited by D. M. Bird, D. E. Varland, and J. J. Negro, 69–74. San Diego: Academic Press.

7. Hager, S. B. 2009. "Human-Related Threats to Urban Raptors." *Journal of Raptor Research* 43:210–26.

8. Botelho, E. S., and P. C. Arrowood. 1996. "Nesting Success of Western Burrowing Owls in Natural and Human-Altered Environments." In *Raptors in Human Landscapes: Adaptations to Built and Cultivated Environments*, edited by D. Bird, D. Varland, and J. J. Negro, 61–68. San Diego: Academic Press.

9. Berardelli, D., M. J. Desmond, and L. Murray. 2010. "Reproductive Success of Burrowing Owls in Urban and Grassland Habitats in Southern New Mexico." *Wilson Journal of Ornithology* 122:51–59.

10. Gehlbach, F. R. 2012. "Eastern Screech-Owl Responses to Suburban Sprawl, Warmer Climate, and Additional Avian Food in Central Texas." *Wilson Journal of Ornithology* 124:630–33.

11. Parker, J. W. 1996. "Urban Ecology of the Mississippi Kite." In *Raptors in Human Landscapes: Adaptations to Built and Cultivated Environments*, edited by D. Bird, D. Varland, and J. J. Negro, 45–52. San Diego: Academic Press.

12. Skipper, B. R. 2013. "Urban Ecology of Mississippi Kites." PhD diss., Texas Tech University.

13. Welch, B. C. 2016. "The Breeding Ecology and Predicted Influence of Climate Change on Urban-Nesting Mississippi Kites." PhD diss., Texas Tech University.

14. Sodhi, N. S., P. C. James, I. G. Warkentin, and L. Oliphant. 1992. "Breeding Ecology of Urban Merlins (*Falco columbarius*)." *Canadian Journal of Zoology* 70:1477–83.

15. Stout, W. E., S. A. Temple, and J. M. Papp. 2006. "Landscape Correlates of Reproductive Success for an Urban–Suburban Red-Tailed Hawk Population." *Journal of Wildlife Management* 70:989–97.

16. Minor, W. F., M. Minor, and M. F. Ingraldi. 1993. "Nesting of Red-Tailed Hawks and Great Horned Owls in a Central New York Urban/Suburban Area." *Journal of Field Ornithology* 64:433–39.

17. Bloom, P. H., M. D. McCrary, and M. J. Gibson. 1993. "Red-Shouldered Hawk Home Range and Habitat Use in Southern California." *Journal of Wildlife Management* 57:258–65.

18. Rottenborn, S. C. 2000. "Nest-Site Selection and Reproductive Success of Urban Red-Shouldered Hawks in Central California." *Journal of Raptor Research* 34:18–25.

19. Dykstra, C. R., J. L. Hays, F. B. Daniel, and M. M. Simon. 2000. "Nest Site Selection and Productivity of Suburban Red-Shouldered Hawks in Southern Ohio." *Condor* 102:401–8.

20. Dykstra, C. R., J. L. Hays, and M. M. Simon. 2009. "Spatial and Temporal Variation in Reproductive Rates of the Red-Shouldered Hawk in Suburban and Rural Ohio." *Condor* 111:177–82.

21. Millsap, B. A., and C. Bear. 2000. "Density and Reproduction of Burrowing Owls along an Urban Development Gradient." *Journal of Wildlife Management* 64:33–41.

22. Millsap, B. A. 2002. "Survival of Florida Burrowing Owls along an Urban Development Gradient." *Journal of Raptor Research* 36:3–10.

23. Hindmarch, S., E. A. Krebs, J. Elliott, and D. J. Green. 2014. "Urban Development Reduces Fledging Success of Barn Owls in British Columbia, Canada." *Condor* 116:507–17.

24. Hindmarch, S., and J. E. Elliott. 2015. "A Specialist in the City: The Diet of Barn Owls along a Rural to Urban Gradient." *Urban Ecosystems* 18:477–88.

25. Salvati, L., L. Ranazzi, and A. Manganaro. 2002. "Habitat Preferences, Breeding Success, and Diet of the Barn Owl (*Tyto alba*) in Rome: Urban versus Rural Territories." *Journal of Raptor Research* 36:224–28.

26. See chapter 7.

27. Boggie, M. A., and R. W. Mannan. 2014. "Examining Seasonal Patterns of Space Use to Gauge How an Accipiter Responds to Urbanization." *Landscape and Urban Planning* 124:34–42.

28. Millsap, B. A., T. F. Breen, and L. M. Phillips. 2013. "Ecology of the Cooper's Hawk in North Florida." *North American Fauna* 78:1–58.

29. Marzluff, J. M., F. R. Gehlbach, and D. A. Manuwal. 1998. "Urban Environments: Influences on Avifauna and Challenges for the Conservationist." In *Avian Conservation: Research and Management*, edited by J. M. Marzluff and R. Sellabanks, 283–92. Washington, DC: Island Press.

30. Boal, C. W. 2001. "Nonrandom Mating and Productivity in Adult and Subadult Cooper's Hawks." *Condor* 103:381–85.

31. Mannan, R. W., R. J. Steidl, and C. W. Boal. 2008. "Identifying Habitat Sinks: A Case Study of Cooper's Hawks in an Urban Environment." *Urban Ecology* 11:141–48.

32. Mannan, R. W., W. A. Estes, and W. J. Matter. 2004. "Movements and Survival of Fledgling Cooper's Hawks in an Urban Environment." *Journal of Raptor Research* 38:26–34.

33. Kauffman, M. J., W. F. Frick, and J. Linthicum. 2003. "Estimation of Habitat Specific Demography and Population Growth for Peregrine Falcons in California." *Ecological Applications* 13:1802–16.

34. Gahbauer, M. A., D. M. Bird, K. E. Clark, T. French, D. W. Brauning, and F. A. McMorris. 2015. "Productivity, Mortality, and Management of Urban Peregrine Falcons in Northeastern North America." *Journal of Wildlife Management* 79:10–19.

35. Dawson, J. W., and R. W. Mannan. 1989. "A Comparison of Two Methods of Estimating Breeding Group Size in Harris' Hawks." *Auk* 106:480–83.

36. Dwyer, J. F., and R. W. Mannan. 2007. "Preventing Raptor Electrocutions in an Urban Environment." *Journal of Raptor Research* 41:259–67.

37. Boal, C. W., and R. W. Mannan. 1999. "Comparative Breeding Ecology of Cooper's Hawks in Urban and Exurban Areas of Southeastern Arizona." *Journal of Wildlife Management* 63:77–84.

38. Battin, J. 2004. "When Good Animals Love Bad Habitats: Ecological Traps and the Conservation of Animal Populations." *Conservation Biology* 18:1482–91.

39. Robertson, B. A., and R. L. Hutto. 2006. "A Framework for Understanding Ecological Traps and an Evaluation of Existing Evidence." *Ecology* 87:1075–85.

40. Isaac, B., R. Cooke, D. Ierodiaconou, and J. White. 2014. "Does Urbanization Have the Potential to Create an Ecological Trap for Powerful Owls (*Ninox strenua*)?" *Biological Conservation* 176:1–11.

41. Estes, W. A., and R. W. Mannan. 2003. "Feeding Behavior of Cooper's Hawks at Urban and Rural Nests in Southeastern Arizona." *Condor* 105:107–16.

42. Steidl, R. J., and B. F. Powell. 2006. "Assessing the Effects of Human Activities on Wildlife." *George Wright Forum* 23:50–58.

CHAPTER 5

Urbanization and Raptors: Trends and Research Approaches

Raylene Cooke, Fiona Hogan, Bronwyn Isaac,

Marian Weaving, and John G. White

URBANIZATION PRESENTS A MAJOR GLOBAL issue for the conservation and survival of many different species. With the increasing footprint of cities and intensification of our use of urban areas, wildlife faces extremely difficult challenges to live there. Understanding how species respond to urban processes and how to design urban landscapes that facilitate species' presences are major emerging research and management priorities. Despite general negative responses to increasing urbanization, some animal taxa, both native and introduced, appear to benefit from urban environments by capitalizing on novel environments and abundant resources.[1] Those that are common in urban systems display particular physical characteristics and ecological traits.[2,3,4] They also frequently display a level of behavioral plasticity or tolerance, adjusting their behavior to interact with, and survive in, urban environments.[5,6] Termed *urban-adaptors*,[7] these species may exhibit altered spatial,[8,9,10] foraging,[11,12] and breeding behaviors,[13] as detailed in chapter 2.

Raptors are top trophic level predators often considered vulnerable to habitat loss,[14,15] and in many instances, their populations and distribution have declined as a direct result of landscape modification by urbanization.[16] Although many raptor species have a negative response to urbanization, there are others that seem to tolerate or even thrive in urban environments. Examples discussed throughout this book include the great horned owl (*Bubo virginianus*), red-shouldered hawk (*Buteo lineatus*), Cooper's hawk (*Accipiter cooperii*), burrowing owl (*Athene cunicularia*), peregrine falcon (*Falco peregrinus*), and powerful owl (*Ninox strenua*).[17,18,19,20,21 22,23] Others, such as the northern goshawk (*Accipiter gentilis*), tolerate urbanization in some locations but not in others.[24]

So what factors influence whether raptors will inhabit an urban area? How do we study animals in highly fragmented landscapes where much of the land is privately owned and access to conduct research can be difficult? How should we assess the levels of disturbance faced by urban raptors, given that they may vary widely within a single animal's home range? There is no doubt that studying raptor ecology is difficult in natural areas, but research focusing on them in urban areas also provides great technical challenges. Yet this research is critically needed. Ultimately, modern approaches to studying ecology—such as spatial modeling, landscape genetics, and remote tracking technologies—may offer the solution to understanding how animals respond to this modern problem.

The tolerance of raptors to urbanization can be related to the environmental attributes and biotic relationships present in urban areas. Stable or abundant prey bases, novel environments, reduced competition, and additional nesting structures—collectively or separately—have been suggested as factors that influence raptor survival and success in urban environments.[24,25] Generalist bird species, insects, small mammals, and a variety of human commensals generally occur at higher densities in urban areas, providing important prey items for many raptors.[24,26,27] As unique landforms, the built environment and urban green space now represent growing and important land-use types for many avian species, including raptors.[28,29] The capacity for raptors to utilize urban landscapes depends on their habitat requirements and thresholds of tolerance to disturbance.[21,30,31] Some raptor species incorporate urban or suburban areas into their home ranges for functions such as foraging and roosting,[32,33,34,35,36] while others actively avoid these environments[37,38] and inhabit only natural areas.

Urban-breeding raptors nest in native[34,39] and nonnative tree species,[30,40] use human-made structures,[20,41,42] use nests built by other avian species,[43,44,45] or exploit domestic features, including window ledges and flowerpots as nest sites.[46] Plentiful breeding sites combined with abundant prey can boost reproductive rates to high levels[46,47] but may be offset by other factors such as increased disease and predation.[48,49,50]

In addition to increased prey abundances, urban environments may provide raptors with novel opportunities to access prey. For example, the proliferation of artificial night light in urban environments has created a unique foraging environment for some raptor species. Peregrine falcons, lesser kestrels (*Falco naumanni*), and burrowing owls (*Athene cunicularia*) have all made adjustments to their foraging behavior in order to exploit prey attracted to these lights.[20,51,52]

Cooper's hawks are regarded as among the most urban-tolerant of North American raptor species.[17,53] Their successful colonization of some urban environments in high numbers has been linked to their capacity to utilize a variety of habitat types.[19,54,55] Along with populations of Cooper's hawks, populations of other raptor species[21,23,56] appear stable and in some areas are expanding their distribution to create new populations.[55,57,58] Even species more traditionally associated with forested areas, such as barred owls (*Strix varia*), are also found in some urban environments.[34,59] However, as urbanization intensifies and human impacts increase, species once considered urban-tolerant may reach a threshold beyond which populations begin to decline.[39,60,61]

Methods Used to Investigate Raptor Responses to Urbanization

Our knowledge of the effects of urbanization on bird communities has advanced significantly in recent years.[62,63,64] In contrast, there is still a lack of information regarding the response of raptors to urbanization.[24] This may be due in part to the low densities, large spatial requirements, and cryptic behaviors of many raptor species.[24,65]

Recent technologies that make use of spatial modeling programs, including FRAGSTATS,[66] Geospatial Modeling Environment (GME),[67] and Geographical Information System (GIS),[68] have become increasingly important tools in raptor research. Using these technologies coupled with data obtained through field techniques such as radiotelemetry and direct observation, researchers have been able to analyze the effects of multiple land-use types on the spatial ecology of a species.[31,69,70] For example, in Melbourne, Australia, researchers applied GIS techniques to produce a predictive map identifying potential breeding sites for the threatened powerful owl.[70] This study has provided critical information on this iconic species, which until recent times was considered to be primarily associated with forested environments.

Advances in tracking technologies have proven a substantial asset for urban raptor research. Traditionally, raptors were tracked by researchers on foot, who used geometry to triangulate the individual bird's position between two or three points. This can result in substantial error in the estimated location, which increases in concert with the distance between the bird and the researcher. New tools, such as transmitters with global positioning system (GPS) loggers, provide a greater volume of data and a higher degree of accuracy compared to traditional tracking methods. Further, these tools are especially helpful in urban settings where access can be severely constrained due to the predominance of privately owned properties. Additionally, satellite technologies have been particularly useful for spatial studies of large-bodied migratory raptors.[37,71]

Harnessing the power of computers to organize and relate large datasets, Species Distribution Modeling (SDM) is useful for analyzing the relationship between a species occurrence (or inferred absence) and a set of ecological and geographical variables.[72,73] In recent years, raptor researchers are increasingly using SDM modeling approaches to identify important habitat variables for the management and conservation of threatened species. One such example is the study of a small population of great grey owls (*Strix nebulosa*), a species identified as highly vulnerable to land-use change and habitat fragmentation, which inhabits the central Sierra Nevada range in California. By using SDM modeling and GIS techniques, researchers were able to identify a critical habitat type, which can now be incorporated into current and future conservation efforts and management plans for the species.[74]

Molecular techniques have also become more widely adopted for raptor research. DNA can be obtained by taking a small blood sample from a bird or, alternatively, by plucking a feather (contour or down) or opportunistically collecting a shed feather.[75] The collection of shed feathers is an attractive sampling option as it negates the need for capturing elusive and/or cryptic species. It is not a panacea, however, as DNA extracted from shed feathers is often degraded (i.e., the yield and quality is poor), which can affect analyses.

Regardless of how the sample is collected, molecular markers that now exist can be used to address ecological questions such as breeding and mating systems; population structure, size, and connectivity; and habitat use and movements across a landscape. Genetic analysis can therefore be a valuable tool for understanding raptor ecology in urbanized ecosystems.

Insights into Urban Raptor Ecology

Raptors, by their very nature, are difficult to study in their natural habitats. However, the built environment, which dominates the urban landscape, presents additional obstacles for raptor research. Even when permission is obtained to access private properties in urban settings, physical structures such as buildings and fences can restrict researcher movements, and artificial light may further hinder the ability to locate and observe nocturnal birds such as owls. Genetic data, used together with tracking technologies and spatial modeling, provide a valuable opportunity to gain insight into the lives of raptors and enable us to investigate how and why these incredible birds, some of which were once thought to be highly selective, are able to survive (and even thrive) in modified, urbanized environments.

Breeding and Dispersing in an Urban Matrix

Natal dispersal is the movement of an individual from their birthplace to their first breeding location.[76] Although dispersal is not without its risks to the individual, these movements are essential for maintaining the genetic integrity of a species. Dispersal within highly modified environments, however, presents an array of challenges for dispersing individuals—including navigation through a matrix of unfavorable environments and avoidance of novel sources of mortality—in their quest to find a territory and mate. If they are successful, the lack of suitable breeding resources may be a further obstacle. Urban environments can therefore reduce fecundity, which can ultimately influence the population dynamics and future viability of a species. The actual effects of urban environments on the breeding and mating systems of raptors, however, remain largely unknown.

Genetic methodologies are increasingly being used to assess the mating systems and population dynamics of birds.[77,78] Extra-pair fertilizations (EPF), for example, are regarded as fairly rare in raptors,[79,80] due to the extent of male parental effort and the large spatial distances between breeding pairs.[81] However, researchers used genetic analyses to determine that the Cooper's hawk, which is common in urban areas in Milwaukee, Wisconsin, has relatively high frequencies of EPFs.[82,83] Genetic analysis can therefore facilitate the study of complex processes such as mating systems and has been used to detect unusual behaviors such as breeding between closely related individuals in urban environments.[84]

Landscape Genetics

Maintaining and improving landscape connectivity has become a priority for the conservation and protection of many species, particularly in fragmented landscapes. Integrating spatial modeling with genetic analysis, also termed *landscape genetics*, offers unprecedented power to assess how landscapes facilitate or impede the movement of individuals. Applications, including GIS and other geospatial modeling tools, enable researchers and conservation planners to improve landscape connectivity.[85,86] Recently developed connectivity models, such as Circuitscape, have distinct advantages over earlier approaches due to their ability to evaluate multiple movement pathways, which can then be applied to ecological processes such as dispersal.[87] The applications of landscape genetic methods to raptor research have primarily addressed raptors associated with more natural landscapes, highlighting demographic changes over time,[88] the influence of landscape features on morphology and genetic structure,[89] and the effect of habitat fragmentation on genetic diversity.[90] New advances in tracking technologies have resulted in GPS loggers becoming much lighter, increasing their application for raptor research.[91] Attaching GPS loggers to adults and/or juveniles provides real-time movements, which can be translated into information about habitat use or dispersal routes.

Conclusions

Maintaining functional populations of urban-dwelling raptors will be a challenge in the coming decades, given the pace and extent of urbanization across the globe. Supporting persistence of current populations in urban environments and enabling new populations to establish themselves will require considerable effort on the part of both researchers and urban planners. Research suggests that some species thrive in urban environments, whereas others decline. An important focus for future research should therefore be to identify species-specific thresholds to urbanization in order to support multispecies occupancy of urban areas. This focus will require collaboration among raptor researchers, with research efforts spanning cities, regions, and countries across the world. This global approach is important, as many threatened raptors use multiple habitats over very large areas. Although this approach may seem daunting, recent technological advances highlighted here will enable research to be undertaken at a larger scale and extent than ever envisaged in the past.

Literature Cited

1. McKinney, M. L. 2002. "Urbanization, Biodiversity, and Conservation." *BioScience* 52:883–90.
2. Møller, A. P. 2009. "Successful City Dwellers: A Comparative Study of the Ecological Characteristics of Urban Birds in the Western Palearctic." *Oecologia* 159:849–58.

3. Prange, S., and S. D. Gehrt. 2004. "Changes in Mesopredator-Community Structure in Response to Urbanization." *Canadian Journal of Zoology* 82:1804–17.

4. Kark, S., A. Iwaniuk, A. Schalimtzek, and E. Banker. 2007. "Living in the City: Can Anyone Become an 'Urban Exploiter'?" *Journal of Biogeography* 34:638–51.

5. Sol, D., O. Lapiedra, and C. González-Lagos. 2013. "Behavioural Adjustments for a Life in the City." *Animal Behaviour* 85:1101–12.

6. Lizée, M.-H., J.-F. Mauffrey, T. Tatoni, and M. Deschamps-Cottin. 2011. "Monitoring Urban Environments on the Basis of Biological Traits." *Ecological Indicators* 11:353–61.

7. McKinney, M. L. 2006. "Urbanization as a Major Cause of Biotic Homogenization." *Biological Conservation* 127:247–60.

8. Gese, E., P. Morey, and S. Gehrt. 2012. "Influence of the Urban Matrix on Space Use of Coyotes in the Chicago Metropolitan Area." *Journal of Ethology* 30:413–25.

9. Hodgson, P., K. French, and R. E. Major. 2007. "Avian Movement across Abrupt Ecological Edges: Differential Responses to Housing Density in an Urban Matrix." *Landscape and Urban Planning* 79:266–72.

10. Dowding, C. V., S. Harris, S. Poulton, and P. J. Baker. 2010. "Nocturnal Ranging Behaviour of Urban Hedgehogs, *Erinaceus europaeus*, in Relation to Risk and Reward." *Animal Behaviour* 80:13–21.

11. Gehlbach, F. R. 2012. "Eastern Screech-Owl Responses to Suburban Sprawl, Warmer Climate, and Additional Avian Food in Central Texas." *Wilson Journal of Ornithology* 124:630–33.

12. Prange S., S. D. Gehrt, and E. P. Wiggers. 2004. "Influences of Anthropogenic Resources on Raccoon (*Procyon lotor*) Movements and Spatial Distribution." *Journal of Mammalogy* 85:483–90.

13. Chamberlain, D. E., A. R. Cannon, M. P. Toms, D. I. Leech, B. J. Hatchwell, and K. J. Gaston. 2009. "Avian Productivity in Urban Landscapes: A Review and Meta-Analysis." *Ibis* 151:1–18.

14. Bildstein, K. L., W. Schelsky, J. Zales, and S. Ellis. 1998. "Conservation Status of Tropical Raptors." *Journal of Raptor Research* 32:3–18.

15. Thiollay, J.-M. 2006. "The Decline of Raptors in West Africa: Long-Term Assessment and the Role of Protected Areas." *Ibis* 148:240–54.

16. Ferrer-Sánchez, Y., and R. Rodríguez-Estrella. 2015. "Man-Made Environments Relationships with Island Raptors: Endemics Do Not Cope with Habitat Changes, the Case of the Island of Cuba." *Biodiversity and Conservation* 24:407–25.

17. Rullman, S., and J. M. Marzluff. 2014. "Raptor Presence along an Urban–Wildland Gradient: Influences of Prey Abundance and Land Cover." *Journal of Raptor Research* 48:257–72.

18. Bloom, P. H., and M. D. McCrary. 1996. "The Urban Buteo: Red-Shouldered Hawks in Southern California." In *Raptors in Human Landscapes: Adaptations to Built and*

Cultivated Environments, edited by D. Bird, D. Varland, and J. J. Negro, 31–39. San Diego: Academic Press.

19. Boal, C. W., and R. W. Mannan. 1999. "Comparative Breeding Ecology of Cooper's Hawks in Urban and Exurban Areas of Southeastern Arizona." *Journal of Wildlife Management* 63:77–84.

20. Botelho, E. S., and P. C. Arrowood. 1996. "Nesting Success of Western Burrowing Owls in Natural and Human-Altered Environments." In *Raptors in Human Landscapes: Adaptations to Built and Cultivated Environments*, edited by D. Bird, D. Varland, and J. J. Negro, 61–68. San Diego: Academic Press.

21. Caballero, I. C., J. M. Bates, M. Hennen, and M. V. Ashley. 2016. "Sex in the City: Breeding Behavior of Urban Peregrine Falcons in the Midwestern US." *PLoS ONE* 11(7): e0159054. doi:10.1371/journal.pone.0159054.

22. Cooke, R., R. Wallis, and A. Webster. 2002. "Urbanization and the Ecology of Powerful Owls *(Ninox strenua)* in Outer Melbourne, Victoria." In *Ecology and Conservation of Owls*, edited by I. Newton, R. Kavanagh, J. Olsen, and I. Taylor, 100–106. Melbourne, Australia: CSIRO Publishing.

23. Rutz, C. 2006. "Home Range Size, Habitat Use, Activity Patterns and Hunting Behaviour of Urban-Breeding Northern Goshawks. *Accipiter gentilis.*" *Ardea* 94:185–202.

24. Chace, J. F., and J. J. Walsh. 2006. "Urban Effects on Native Avifauna: A Review." *Landscape and Urban Planning* 74:46–69.

25. Gehlbach, F. R. 1996. "Eastern Screech Owls in Suburbia: A Model of Raptor Urbanization." In *Raptors in Human Landscapes: Adaptations to Built and Cultivated Environments*, edited by D. M. Bird, D. E. Varland, and J. J. Negro, 69–74. San Diego: Academic Press.

26. Green, R. 1984. "Native and Exotic Birds in a Suburban Habitat." *Wildlife Research* 11:181–90.

27. Savard, J. P. L., P. Clergeau, and G. Mennechez. 2000. "Biodiversity Concepts and Urban Ecosystems." *Landscape and Urban Planning* 48:131–42.

28. Sandström, U. G., P. Angelstam, and G. Mikusiński. 2006. "Ecological Diversity of Birds in Relation to the Structure of Urban Green Space." *Landscape and Urban Planning* 77:39–53.

29. Morrison, J. L., I. G. W. Gottlieb, and K. E. Pias. 2016. "Spatial Distribution and the Value of Green Spaces for Urban Red-Tailed Hawks." *Urban Ecosystems* 19:1373–88.

30. Boal, C. W., and R. W. Mannan. 1998. "Nest-Site Selection by Cooper's Hawks in an Urban Environment." *Journal of Wildlife Management* 62:864–71.

31. Sumasgutner, P., E. Nemeth, G. Tebb, H. W. Krenn, and A. Gamauf. 2014. "Hard Times in the City—Attractive Nest Sites but Insufficient Food Supply Lead to Low Reproduction Rates in a Bird of Prey." *Frontiers in Zoology* 11:48.

32. Bennett, J. R., and P. H. Bloom. 2005. "Home Range and Habitat Use by Great Horned Owls (*Bubo virginianus*) in Southern California." *Journal of Raptor Research* 39:119–26.

33. De Giacomo, U., and G. Guerrieri. 2008. "The Feeding Behavior of the Black Kite (*Milvus migrans*) in the Rubbish Dump of Rome." *Journal of Raptor Research* 42:110–18.

34. Dykstra, C. R., M. M. Simon, F. B. Daniel, and J. L. Hays. 2012. "Habitats of Suburban Barred Owls (*Strix varia*) and Red-Shouldered Hawks (*Buteo lineatus*) in Southwestern Ohio." *Journal of Raptor Research* 46:190–200.

35. Sodhi, N. S., and L. W. Oliphant. 1992. "Hunting Ranges and Habitat Use and Selection of Urban-Breeding Merlins." *Condor* 94:743–49.

36. Hindmarch, S., and J. E. Elliott. 2015. "A Specialist in the City: The Diet of Barn Owls along a Rural to Urban Gradient." *Urban Ecosystems* 18:477–88.

37. Domenech, R., B. E. Bedrosian, R. H. Crandall, and V. A. Slabe. 2015. "Space Use and Habitat Selection by Adult Migrant Golden Eagles Wintering in the Western United States." *Journal of Raptor Research* 49:429–40.

38. Bosakowski, T., and D. G. Smith. 1997. "Distribution and Species Richness of a Forest Raptor Community in Relation to Urbanization." *Journal of Raptor Research* 31:26–33.

39. Dykstra, C. R., J. L. Hays, F. B. Daniel, and M. M. Simon. 2000. "Nest Site Selection and Productivity of Suburban Red-Shouldered Hawks in Southern Ohio." *Condor* 102:401–8.

40. Skipper, B. R. 2013. "Urban Ecology of Mississippi Kites." PhD diss., Texas Tech University.

41. Martell, M. S., J. V. Englund, and H. B. Tordoff. 2002. "An Urban Osprey Population Established by Translocation." *Journal of Raptor Research* 36:91–96.

42. Stout, W. E., R. K. Anderson, and J. M. Papp. 1996. "Red-Tailed Hawks Nesting on Human-Made and Natural Structures in Southeast Wisconsin." In *Raptors in Human Landscapes: Adaptations to Built and Cultivated Environments*, edited by D. Bird, D. Varland, and J. J. Negro, 77–86. San Diego: Academic Press.

43. Lövy, M., and J. Riegert. 2013. "Home Range and Land Use of Urban Long-Eared Owls." *Condor* 115:551–57.

44. Smith, D. G., T. Bosakowski, and A. Devine. 1999. "Nest Site Selection by Urban and Rural Great Horned Owls in the Northeast." *Journal of Field Ornithology* 70:535–42.

45. Warkentin, I. G., and P. C. James. 1988. "Nest-Site Selection by Urban Merlins." *Condor* 90:734–38.

46. Charter, M., I. Izhaki, A. Bouskila, Y. Leshem, and V. Penteriani. 2007. "Breeding Success of the Eurasian Kestrel (*Falco tinnunculus*) Nesting on Buildings in Israel." *Journal of Raptor Research* 41:139–43.

47. Solonen, T. 2008. "Larger Broods in the Northern Goshawk *Accipiter gentilis* near Urban Areas in Southern Finland." *Ornis Fennica* 85:118–25.

48. Hindmarch, S., E. A. Krebs, J. Elliott, and D. J. Green. 2014. "Urban Development Reduces Fledging Success of Barn Owls in British Columbia, Canada." *Condor* 116:507–17.

49. Boal, C. W., R. W. Mannan, and K. S. Hudelson. 1998. "*Trichomoniasis* in Cooper's Hawks from Arizona." *Journal of Wildlife Diseases* 34:590–93.

50. Miller, S. J., C. R. Dykstra, M. M. Simon, J. L. Hays, and J. C. Bednarz. 2015. "Causes of Mortality and Failure at Suburban Red-Shouldered Hawk (*Buteo lineatus*) Nests." *Journal of Raptor Research* 49:152–60.

51. DeCandido, R., and D. Allen. 2006. "Nocturnal Hunting by Peregrine Falcons at the Empire State Building, New York City." *Wilson Journal of Ornithology* 118:53–58.

52. Negro, J. J., J. Bustamante, C. Melguizo, J. L. Ruiz, and J. M. Grande. 2000. "Nocturnal Activity of Lesser Kestrels under Artificial Lighting Conditions in Seville, Spain." *Journal of Raptor Research* 34:327–29.

53. Boggie, M. A., and R. W. Mannan. 2014. "Examining Seasonal Patterns of Space Use to Gauge How an Accipiter Responds to Urbanization." *Landscape and Urban Planning* 124:34–42.

54. Roth, T. C., II., and S. L. Lima. 2003. "Hunting Behavior and Diet of Cooper's Hawks: An Urban View of the Small-Bird-in-Winter Paradigm." *Condor* 105:474–83.

55. Stout, W. E., and R. N. Rosenfield. 2010. "Colonization, Growth, and Density of a Pioneer Cooper's Hawk Population in a Large Metropolitan Environment." *Journal of Raptor Research* 44:255–67.

56. Stout, W. E., S. A. Temple, and J. M. Papp. 2006. "Landscape Correlates of Reproductive Success for an Urban-Suburban Red-Tailed Hawk Population." *Journal of Wildlife Management* 70:989–97.

57. Cade, T. J., M. Martell, P. Redig, G. Septon, and H. Tordoff. 1996. "Peregrine Falcons in Urban North America." In *Raptors in Human Landscapes: Adaptations to Built and Cultivated Environments*, edited by D. Bird, D. Varland, and J. J. Negro, 3–13. San Diego: Academic Press.

58. Millsap, B., T. I. M. Breen, E. McConnell, T. Steffer, L. Phillips, N. Douglass, S. Taylor, and C. W. Boal. 2004. "Comparative Fecundity and Survival of Bald Eagles Fledged from Suburban and Rural Natal Areas in Florida." *Journal of Wildlife Management* 68:1018–31.

59. Livezey, K. B. 2007. "Barred Owl Habitat and Prey: A Review and Synthesis of the Literature." *Journal of Raptor Research* 41:177–201.

60. Strasser, E. H., and J. A. Heath. 2013. "Reproductive Failure of a Human-Tolerant Species, the American Kestrel, Is Associated with Stress and Human Disturbance." *Journal of Applied Ecology* 50:912–19.

61. Nagy, C. M., and R. F. Rockwell. 2013. "Occupancy Patterns of *Megascops asio* in Urban Parks of New York City and Southern Westchester County, NY, USA." *Journal of Natural History* 47:2135–49.

62. Beissinger, S. R., and D. R. Osborne. 1982. "Effects of Urbanization on Avian Community Organization." *Condor* 84:75–83.

63. Crooks, K. R., A. V. Suarez, and D. T. Bolger. 2004. "Avian Assemblages along a Gradient of Urbanization in a Highly Fragmented Landscape." *Biological Conservation* 115:451–62.

64. Sol, D., C. González-Lagos, D. Moreira, J. Maspons, and O. Lapiedra. 2014. "Urbanisation Tolerance and the Loss of Avian Diversity." *Ecology Letters* 17:942–50.

65. Isaac, B., J. White, D. Ierodiaconou, and R. Cooke. 2013. "Response of a Cryptic Apex Predator to a Complete Urban to Forest Gradient." *Wildlife Research* 40:427–36.

66. McGarigal, K., S. A. Cushman, and E. Ene. 2012. *FRAGSTATS v4: Spatial Pattern Analysis Program for Categorical and Continuous Maps.* Amherst: University of Massachusetts.

67. Beyer, H. L. 2012. *Geospatial Modelling Environment.* 0.7.1.0 ed. Accessed April 28, 2017. http://www.spatialecology.com/gme%3e.

68. ESRI. 2011. *ArcGIS Desktop: Release 10.* Redlands: Environmental Systems Research Institute.

69. Boggie, M. A., R. W. Mannan, and C. Wissler. 2015. "Perennial Pair Bonds in an Accipiter: A Behavioral Response to an Urbanized Landscape?" *Journal of Raptor Research* 49:458–70.

70. Isaac, B., R. Cooke, D. Simmons, and F. Hogan. 2008. "Predictive Mapping of Powerful Owl (*Ninox strenua*) Breeding Sites Using Geographical Information Systems (GIS) in Urban Melbourne, Australia." *Landscape and Urban Planning* 84:212–18.

71. Vidal-Mateo, J., U. Mellone, P. López-López, J. De La Puente, C. García-Ripollés, A. Bermejo, and V. Urios. 2016. "Wind Effects on the Migration Routes of Trans-Saharan Soaring Raptors: Geographical, Seasonal, and Interspecific Variation." *Current Zoology* 62:89–97.

72. Guisan, A., and N. E. Zimmermann. 2000. "Predictive Habitat Distribution Models in Ecology." *Ecological Modelling* 135:147–86.

73. Miller, J. 2010. "Species Distribution Modeling." *Geography Compass* 4:490–509.

74. Jepsen, E. P. B., J. J. Keane, and H. B. Ernest. 2011. "Winter Distribution and Conservation Status of the Sierra Nevada Great Gray Owl." *Journal of Wildlife Management* 75:1678–87.

75. Hogan, F. E., R. Cooke, C. P. Burridge, and J. A. Norman. 2008. "Optimizing the Use of Shed Feathers for Genetic Analysis." *Molecular Ecology Resources* 8:561–67.

76. Endler, J. 1977. *Geographic Variation, Speciation and Clines.* Princeton: Princeton University Press.

77. Van Den Bussche, R. A., S. A. Harmon, R. J. Baker, A. L. Bryan, J. A. Rodgers, M. J. Harris, and I. L. Brisbin. 1999. "Low Levels of Genetic Variability in North American Populations of the Wood Stork (*Mycteria americana*)." *Auk* 116:1083–92.

78. Rodríguez-Muñoz, R., P. M. Mirol, G. Segelbacher, A. Fernández, and T. Tregenza. 2007. "Genetic Differentiation of an Endangered Capercaillie (*Tetrao urogallus*) Population at the Southern Edge of the Species Range." *Conservation Genetics* 8:659–70.

79. Alcaide, M., J. J. Negro, D. Serrano, J. L. Tella, and C. Rodríguez. 2005. "Extra-Pair Paternity in the Lesser Kestrel *Falco naumanni*: A Re-Evaluation Using Microsatellite Markers." *Ibis* 147:608–11.

80. Rodríguez-Martínez, S., M. Carrete, S. Roques, N. Rebolo-Ifrán, and J. L. Tella. 2014. "High Urban Breeding Densities Do Not Disrupt Genetic Monogamy in a Bird Species." *PLoS ONE* 9: e91314.

81. Müller, W., J. Epplen, and T. Lubjuhn. 2001. "Genetic Paternity Analyses in Little Owls (*Athene noctua*): Does the High Rate of Paternal Care Select against Extra-Pair Young?" *Journal für Ornithologie* 142:195–203.

82. Rosenfield, R. N., T. G. Driscoll, R. P. Franckowiak, L. J. Rosenfield, B. L. Sloss, M. A. Bozek, and J. R. Belthoff. 2007. "Genetic Analysis Confirms First Record of Polygyny in Cooper's Hawks." *Journal of Raptor Research* 41:230–34.

83. Rosenfield, R. N., S. A. Sonsthagen, W. E. Stout, and S. L. Talbot. 2015. "High Frequency of Extra-Pair Paternity in an Urban Population of Cooper's Hawks." *Journal of Field Ornithology* 86:144–52.

84. Hogan, F. E., and R. Cooke. 2010. "Insights into the Breeding Behaviour and Dispersal of the Powerful Owl (*Ninox strenua*) through the Collection of Shed Feathers." *Emu* 110:178–84.

85. Tang, G. S. Y., K. R. Sadanandan, and F. E. Rheindt. 2016. "Population Genetics of the Olive-Winged Bulbul (*Pycnonotus plumosus*) in a Tropical Urban-Fragmented Landscape." *Ecology and Evolution* 6:78–90.

86. Sacks, B. N., J. L. Brazeal, and J. C. Lewis. 2016. "Landscape Genetics of the Nonnative Red Fox of California." *Ecology and Evolution* 6:4775–91.

87. McRae, B. H., B. G. Dickson, T. H. Keitt, and V. B. Shah. 2008. "Using Circuit Theory to Model Connectivity in Ecology, Evolution, and Conservation." *Ecology* 89:2712–24.

88. Garcia, J. T., F. Alda, J. Terraube, F. Mougeot, A. Sternalski, V. Bretagnolle, and B. Arroyo. 2011. "Demographic History, Genetic Structure and Gene Flow in a Steppe-Associated Raptor Species." *BMC Evolutionary Biology* 11:333.

89. Hull, J. M., A. C. Hull, B. N. Sacks, J. P. Smith, and H. B. Ernest. 2008. "Landscape Characteristics Influence Morphological and Genetic Differentiation in a Widespread Raptor (*Buteo jamaicensis*)." *Molecular Ecology* 17:810–24.

90. Barrowclough, G. F., J. G. Groth, L. A. Mertz, and R. J. Gutiérrez. 2006. "Genetic Structure of Mexican Spotted Owl (*Strix occidentalis lucida*) Populations in a Fragmented Landscape." *Auk* 123:1090–102.

91. Bradsworth, N., J. G. White, B. Isaac, and R. Cooke. 2017. "Species Distribution Models Derived from Citizen Science Data Predict the Fine Scale Movements of Owls in an Urbanizing Landscape." *Biological Conservation* 213:27–35.

PART II

Urban Raptors

CHAPTER 6

Mississippi Kites: Elegance Aloft

Ben R. Skipper

MISSISSIPPI KITES (*ICTINIA MISSISSIPPIENSIS*) ARE arguably one of the most abundant raptors that breed in urban and suburban areas within North America (color plate 1). Each summer, urban residents of the southern Great Plains can look skyward and see a dazzling number of these long-winged raptors soaring effortlessly overhead and swooping and diving to catch their aerial insect prey. In urban areas of the southern Great Plains, where kites are abundant, up to five pairs have been documented nesting within the area of a single city block. Such high densities are likely products of both Mississippi kite natural history and the environmental history of the southern Great Plains. Although Mississippi kites appear well-adapted to the southern Great Plains, their past distribution within this ecoregion was likely much more limited prior to settlement by persons of European descent.

In the early 1900s, populations of Mississippi kites in the southeastern United States had declined across their range, and habitat loss was a key factor driving population declines.[1,2] These declines appear to have been short-lived, however, with populations rebounding by midcentury. During the early 1900s, Great Plains kite populations appeared to be stable or expanding.[1] Numerous kites inhabited western Oklahoma, where low-growing shinnery oak (*Quercus havardii*) is abundant.[3] These observations suggested that kites had expanded outward from riparian forests and now colonized upland brushlands and shrub-lands. Around midcentury, the maturation of trees planted prior to and after the Dust Bowl apparently shifted the geographic distribution of kites further to the west.[4] Kite researcher James Parker speculated that the planting of trees in the prairie landscapes of Kansas, Oklahoma, and Texas was a key factor leading to colonization of urban areas by Mississippi kites and hypothesized that the concurrent loss of riparian woodlands and the maturation of shelterbelt plantings prompted kites to settle closer to people, all the while selecting for greater tolerance of people by kites.[5]

By the 1960s, urban-nesting kites were common in southwestern Kansas and from western Oklahoma to the Texas Panhandle. The close proximity of farmland shelterbelts to cities and urban areas likely provided the opportunity for kites to make the leap from rural breeding to urban breeding.[5] From the 1970s to today, kites have continued to occupy urban areas in the southern Great Plains and have achieved remarkable breeding densities. Although Great Plains populations of urban kites are the more studied, kites in the eastern populations also increasingly breed in urban areas, although breeding densities in eastern cities appear lower than those in the southern Great Plains.

Behavior and Ecology

Habitat

At a continental scale, Mississippi kite habitat requirements and nest-site selection appear to be highly plastic. Kites in exurban areas of the east generally select large tracts of bottomland hardwood forests[2,6] with open areas for foraging nearby.[5] In exurban areas of the southern Great Plains, nesting habitat is variable and includes stunted oak and mesquite (*Prosopis* spp.) savannas, windbreaks planted in agricultural areas, and riparian woodlands populated with tall cottonwoods (*Populus* spp.) and other hardwoods. In Arizona and central and western New Mexico, kites are generally limited to riparian woodlands where mature cottonwood and saltcedar (*Tamarix* spp.) grow. Mississippi kites nest along the exurban-urban gradient from low-density housing near city peripheries to high-density suburban housing, as long as trees of adequate stature are present. Kites appear to avoid nesting in urban centers, although it is not known if this is due to lack of trees, lack of prey, or human disturbance.

At a fine scale, habitat selection by eastern exurban kites appears relatively consistent within and across study areas. Kites in eastern exurban populations nest at great heights in tall (24–43 m), large-diameter trees in mature forests.[7,8,9] Within exurban areas of the southern Great Plains, fine-scale habitat selection appears more relaxed, and kites tolerate a wider range of nest trees and nesting areas, from mesquite shrubs to mature cottonwoods.[10] Consequentially, nest heights in exurban areas of the southern Great Plains range from 5 to 17 m.[10] In a study of selection of windbreaks for nesting, kites demonstrated a weak but inconsistent selection for larger windbreaks surrounded by more vegetation that is native.[11] Exurban-nesting kites along riparian woodlands exhibited weak selection for nest trees taller than randomly selected trees.[12] In the southwestern United States (Arizona and western New Mexico), kites consistently selected nest trees greater than 15 m in height and preferred to place their nests in clumped groves rather than isolated trees.[13]

In urban areas of the southern Great Plains, fine-scale habitat selection by urban Mississippi kites seems to mirror that of exurban kites in that they appear to use a wide variety of nest tree species with variable physical characteristics. There was no clear pattern of nest-site selection among an urban population in central Texas over two breeding seasons, except that kites built 72 percent of nests in pecan trees (*Carya illinoinensis*).[14] In a four-year study of urban-nesting kites in the Texas southern High Plains region, Mississippi kites generally selected nest trees with greater height and trunk diameter than random, but there were also many instances of kites using comparatively short and narrow-diameter trees.[12]

DIET

Mississippi kites are typically considered insectivorous, although this characterization grossly oversimplifies their diet. The characterizations of kites as insectivores stems from some of the earliest studies of the species[3,15] and more recent observations of foraging adults and studies of food provisioned to young. In addition to insects, prey items taken by kites include small birds, small mammals, reptiles, and amphibians.[16] There is some evidence that diet may vary between urban and exurban areas and that changing environmental conditions within urban or exurban areas may force kites to shift their diet.[16,17]

Of the prey items taken by kites, none are consistently more prevalent, or perhaps more important, than cicadas (family Cicadidae). The importance of cicadas to Mississippi kites is well known. Upon observing foraging kites and the stomach contents of three collected specimens, Alexander Wilson concluded in 1811 that cicadas represented the primary food of Mississippi kites.[15] Cicadas are among the principle food items fed to young,[14,16] and their abundance may allow kites to achieve the high nesting densities observed in the Great Plains, according to some researchers.[15] In addition to cicadas, other insect taxa may be regionally or seasonally important for Mississippi kites. Cicadas were important during the early breeding season in exurban areas in Arizona,[13] but they were a more important late-season food in exurban Arkansas, where dragonflies dominated in the early season.[17] In contrast, two studies of urban kites indicated that cicadas were important foods throughout the entire nesting season.[14,16]

Mississippi kites apparently have the ability to capitalize on locally abundant prey and a great potential for dietary plasticity. In Arizona, one pair of exurban-nesting kites exploited a nearby population of bats (*Pipistrellus hesperus*).[13] In Arkansas, water level in bottomland hardwood forest was an excellent predictor of diet; in dry years, kites shifted their foraging toward cicadas, whereas in wet years, the diet shifted toward dragonflies (family Odonata).[17] In an urban population, kites also seemed to cope with decreased availability of their primary prey: they provisioned their young with a greater proportion of avian prey during a severe drought year compared to years with typical precipitation.[16]

Other Behaviors

Mississippi kites sometimes reuse existing nests in both urban and exurban areas; however, reuse rates are inconsistent. In exurban areas, 13–50 percent of nests are reused,[5,6,10] whereas in urban areas, reuse of nests ranges from 0 to 45 percent.[14,18] Although explanations of nest-reuse patterns include both adaptive (e.g., time and energy savings) and maladaptive (e.g., higher ectoparasite loads) effects on settling adults, it is likely that nest permanence plays *a* major, if not *the* major, role in determining nest reuse in Mississippi kites. Nests constructed by Mississippi kites usually appear flimsy and are often a loose aggregation of sticks placed in a fork of a limb, in a crotch of a tree, or on a flat portion of a limb.[19] Thus Mississippi kite nests do not persist for multiple years in areas with high winds or severe storms.

Despite the potential for nests to be lost between breeding seasons, kites appear to display some fidelity to localized breeding areas. Although definitions of what constitutes reoccupancy of nesting areas differ, this behavior appears common in both urban and exurban areas. Kites reused 60–68 percent of nests in exurban populations along riparian forests in Arizona[13] and reportedly had high site tenacity in exurban areas of the Great Plains.[10] Similarly, there was a tendency for urban-breeding kites in central Texas to reoccupy previously used nesting areas.[14] Among a banded population of Mississippi kites in urban Lubbock, Texas, year-to-year individual site fidelity to nesting areas was 62 percent, and interannual pair fidelity was 73 percent over four years.[12]

Population Ecology

Reproduction

Reproduction of Mississippi kites differs in several respects between urban and exurban populations. In both environments, kites attempt a single nesting effort each year, although pairs whose first effort fails early may make a second reproductive effort.[20] For both exurban and urban populations, pair bonds appear already formed on arrival at the breeding grounds.[3,5,12,18] The pairs begin copulation and nest-building shortly after arrival. The number of eggs laid is small in both urban and exurban populations of Mississippi kites—typically one to two eggs and rarely three eggs.[10,18] Great Plains populations may be more likely to lay two eggs than kites breeding in eastern populations, the difference possibly attributable to food availability.[10]

Raptors typically nest earlier in urban settings compared to exurban areas.[21] Interestingly, the median hatch date was earlier for Mississippi kites in exurban populations compared with urban populations in West Texas.[22] Additionally, urban hatch dates varied annually, while exurban hatch dates did not.[22] Increased predation pressure in exurban areas may select for earlier nest initiation to allow time for potential renesting if the first effort is lost.[22] The incubation lengths and the nestling-rearing periods are similar between urban and exurban populations, and behaviors of male and female kites around the nests appear, for the most part, similar between urban and exurban areas.

The most striking difference in reproduction between urban and exurban populations is the greater nesting success of urban-nesting Mississippi kites. Across five studies of urban kites from 1977 to 2016, nesting success (i.e., percentage of nests where at least one young fledged) ranged from 33 to 100 percent with a mean of 71.3 percent (± 19.7 [SD]).[5,12,14,18,21] Nesting success in exurban populations ranged from 14 to 50 percent.[5,7,8,9,12,23] While many of the nesting success figures reported herein focus on urban or exurban birds exclusively and the samples are widely spaced geographically and across time, one study[12] examined urban and exurban nesting success both concurrently and in the same geographic space. The results were unequivocal; Mississippi kites find greater nesting success within cities than in undeveloped areas.

Causes of nesting failure among urban and exurban kites are many, but studies focusing on urban-breeding kites more often cite weather and climate as causes of nest failures rather than predators,[5,12,14,20,22] whereas studies of exurban kites more frequently cite predation as a primary cause of nest failure.[5,8,9,17] Undoubtedly, some of the variability in attribution of cause of nest failure is due to differing methodologies; many researchers have used nest cameras to monitor exurban nests but not urban nests. Still, it appears that predation on eggs or nestlings is reduced in urban environments. Although urban environments are not free of predators, some predators of exurban kite nests (e.g., tree-climbing snakes)[8,9] may not be present, or are only present in lower numbers.[24]

Survival

The principle cause of mortality for adult Mississippi kites is unknown. Anecdotal evidence suggests that collisions with automobiles and utility or guy lines do occur, but the prevalence of these mortality sources is likely low.[14] Direct predation by nocturnal predators (raccoons [*Procyon lotor*] and great-horned owls [*Bubo virginianus*]) on adult kites is known in exurban areas, but only for birds attending nests.[10] Although data are lacking, it is assumed that urban Mississippi kites face similar predation pressures as kites nesting in exurban areas but possibly at a lower rate. In both urban and exurban areas, predation of young is likely greater than that of adults.

In urban areas, humans sometimes directly persecute Mississippi kites. For example, in 1978, 28 kites were shot in Ashland, Kansas, because they (or some of them) were diving at people.[20] Although a direct mortality event of this magnitude has not been reported since, wildlife rehabilitation centers still see admission of adult Mississippi kites found with broken or injured wings in urban parks, suggesting that some persecution by humans continues.

Anecdotal evidence of high survivorship in kites has, as with other raptors, been suggested based on reoccupancy of nesting areas year after year.[20] The current longevity record listed by the US Geological Survey Bird Banding Laboratory (BBL) is 11 years. Two other records reported by the BBL document kites that survived to 7 years and 8 years of age after banding in their first year. It is unknown whether these individuals were from urban or exurban populations; however, a nestling kite found in an undeveloped but heavily used park survived to 6 years after being rescued, rehabilitated, and released by a wildlife rehabilitation center.[25] The only annual survival rate reported for adult Mississippi kites is for an urban population in Lubbock, Texas.[12] There, Mississippi kites were estimated to have a 75 percent (59–86, 95 percent confidence interval [CI]) survival rate from one year to the next. This survival estimate is within the range that would be expected for a raptor of this body size.[12]

DENSITY

Populations of nesting kites in the Great Plains of the United States achieve remarkably high nesting densities, leading some authors to characterize Mississippi kites of the Great Plains as colonial or semicolonial.[19,20] In Great Plains populations of Mississippi kites, nearest-neighbor distance is variable but is generally less than that of other species of raptors. Interestingly, studies of nearest-neighbor distance from the Great Plains have revealed relatively consistent distances between nests. Mississippi kites in Kansas nest at a mean neighbor distance of 115 m,[5] with some nests as close as 30 m from each other in agricultural shelterbelts.[10] In the Texas Panhandle, kites nesting along a riparian forest averaged 146–242 m between nests across three years, with a minimum neighbor distance of 40 m.[12] Further west in Arizona, kites nested farther apart (mean of 550 m), with a minimum distance between nests of 125 m.[13]

In urban areas, Mississippi kites nest as close to one another as in exurban areas. In southwestern Kansas, mean distance between Mississippi kite nests was 111 m,[5] whereas in Lubbock, Texas, nearest neighbors averaged 124–203 m apart across four years of observation.[12] In Lubbock, two pairs of kites settled within 30 m of each other with no antagonistic interaction recorded. Although Mississippi kites will nest within close proximity to one another, nesting near other kites is not obligatory. There is documentation of several instances of isolated trees within open spaces, each hosting a single nesting pair.[12]

The divorce of food resources from the nesting territory likely contributes to the high nesting density of some kite populations. Cicadas and other flying insects are an important food for Mississippi kites. Given the vertical distribution of these food items within the air column, kites would little improve their access to these primary food resources by vigorously defending their territories against other kites.

Community Ecology

INTRA- AND INTERSPECIFIC INTERACTIONS

Given the high nesting densities of urban populations (see above), it is not surprising that antagonistic interactions between kites are rarely described. Mississippi kites appear to be very tolerant of neighboring kites and even congregate into roosting areas during the breeding season (figure 6.1). The few antagonistic interactions that have been recorded have generally involved adults driving year-lings out of close proximity (<20 m) of a nest.[14] Additionally, extra-pair copula-tions have been observed in both urban and exurban populations; in both the urban and exurban situations, the territorial male quickly attempted to drive the rival male from the nesting territory.[26,27] The tolerance for conspecifics is likely a result of diet. Cicadas, beetles, and other flying insects are likely distrib-uted patchily throughout the air column, and the location of nesting territories does not necessarily relate to food resources in the urban environment. Thus the presence of conspecifics near the nesting area does not constitute a threat to the territorial pair's access to the food supply.

Urban-nesting Mississippi kites respond vigorously to the presence of would-be predators both near and away from their nests. Kites actively mob many larger diurnal raptors, larger owls, and crows and ravens.[20] Although these species could certainly be considered a threat to eggs, nestlings, or perhaps adults, kites also mob nonthreatening species such as turkey vultures (*Cathartes aura*). Interestingly, the high nesting densities of urban kite populations often result in

Figure 6.1. Mississippi kites are tolerant of neighboring kites and even congregate into roosting areas during the breeding season; there are six individuals in this one picture. Antagonistic interactions generally only occur in close proximity (<20 m) of a nest.

group mobbing of predators. Flocks of more than 20 kites circling and diving at potential predators are not uncommon where kite nesting density is high.

INTERACTIONS WITH HUMANS

Human-wildlife interactions in urban environments are often complex and, in many ways, a product of individual experiences, a topic covered in greater depth in part 3. Where birds of prey occur in urban landscapes, many human residents are afforded an opportunity they would not otherwise have: the chance to closely observe predator behavior. For many, such encounters can be enriching and can lead to a deep fondness and sense of ownership, but when interactions are negative, attitudes may quickly turn against urban wildlife.[28,29]

The juxtaposition of suitable breeding areas for Mississippi kites and areas that humans frequent can set off confrontations between the two species. Wherever kites breed in urban areas, each spring will undoubtedly bring some kites and residents into conflict over a backyard, a city park, or another green space. In such interactions, kites may repeatedly dive at people who, often unknowingly, have ventured too close to nests (figure 6.2). Kites rarely make contact with pedestrians, and the diving behaviors often cease once kites have successfully driven off the intruder.[30,31] For most such encounters, pedestrians quickly leave the nesting area and harbor no ill will toward the offending kite or the population as a whole. In other cases, persons may feel threatened by diving kites and may retaliate by striking them with objects (e.g., umbrellas, handbags, golf clubs, etc.), thus injuring the birds. Moreover, because kites are protected under both US federal law and state law, persons who injure diving kites are subject to prosecution, a situation that reinforces negative perceptions about the species as a whole and as urban raptors in general.

Factors influencing the onset of aggressive nest defense in Mississippi kites are currently unknown. There was no apparent relationship between levels of human activity and diving behaviors at a golf course in Clovis, New Mexico, and several other studies have supported the lack of relationship between the amount of pedestrian traffic and aggression.[30] For example, aggressive nest defense by kites occurred in an exurban population with little to no human presence.[7] A study to explicitly test conditions leading to aggression revealed variable responses of kites to model pedestrians in three habitats with different levels of foot traffic, although aggressive diving behaviors were present in all three habitats.[12] Furthermore, there were no consistent relationships between nest-site characteristics and diving behaviors in urban areas.[12] Anecdotally,

Figure 6.2. **When urbanites venture too close to the nest of some Mississippi kites, adult kites may respond aggressively and attempt to drive the pedestrian away with repeated swoops and dives. Although contact with the pedestrian is rare, the encounter may leave a lasting impression on the pedestrian.**

direct negative experience with people may play some role in the onset of diving behaviors; at two nests in Texas, harassment by humans resulted in the onset of aggressive diving behaviors by kites (C. Boal, pers. comm.).

Much of the public erroneously assigns aggressive nest defense behaviors to all nesting kites. Careful quantification, however, reveals that the prevalence is lower than estimated by the public. In an urban population in San Angelo, Texas, 47 percent of nesting kite pairs dove at pedestrians.[14] Elsewhere, diving occurred at less than 20 percent of nests[31] and in only 12 percent of pedestrian trials in a study directly evaluating the behavior.[12]

Conservation and Management

Following the unlawful shooting of kites in Ashland, Kansas, there was intense interest in developing management tools to minimize human-kite conflict. Removal and translocation of young from aggressive pairs to passive pairs of kites successfully reduced aggressive encounters with humans.[5] This technique is time- and labor-intensive and is most effective during the nestling stage, as the removing of eggs may not be useful due to the renesting potential of the aggressive pair.[30] The translocation of young from one nest to another raises some ethical concerns for the foster parents and their young: the addition of young to an existing brood may lead to additional stress on the natural young, the foster parents, or both. Further, evidence of successful translocations in which translocated young and their natural nest-mates fledge successfully is scarce, because survival of those young or the adults who tended them has not been studied.

Dissuading kites from using nest areas that have high human traffic, and therefore a greater potential for negative interactions, offers a proactive avenue for managing aggressive behaviors. As yet, there are no known consistent habitat features that are predictive of aggression. However, the high site fidelity displayed by urban breeding kites does allow managers to predict where problem areas may be from year to year. The placement of kite effigies in nest trees prior to the arrival of kites may deter settlement in particular areas.[30] In eastern New Mexico, effigies were moderately successful at reducing nest tree reuse by kites (10 percent with effigies, 35–45 percent without effigies).[30] Effectiveness of effigies may be dependent on their appearance to live kites; where effigies become damaged or are moved from lifelike postures, kites appear to ignore the models and settle nearby.

Kites might also be dissuaded from nesting in areas by modifying or removing trees. Although this technique has not been attempted experimentally with kites, experiments with other birds[32] and anecdotal observations after tree or limb removal for landscaping purposes or due to storm damage have shown that this encouraged nesting kites to move. Given the high site fidelity of kites, resettlement distances after the removal of the nest or trimming of the nest tree may be minimal. Additionally, encouraging kites to vacate one area for another may have unintended effects, especially if kites elect to nest in an area where they would be less tolerated by humans. Finally, for removal of limbs or trees to be legal, nesting birds (including birds other than kites) must not have eggs or young in the nest at the time of modification.

Awareness campaigns and public education are likely the most cost-effective means for minimizing human-kite interactions. Signs that warn pedestrians of the potential for diving by kites can be effective at changing human activity patterns and thus reduce negative interactions (figure 6.3). Further, signs are low-cost and can be rapidly deployed in problem areas. Lastly, educating the public about why kites dive at people may effect a change in attitude and bring about an understanding that the nuisance behaviors are only temporary.

Human-raptor conflicts near urban nests are by no means restricted to Mississippi kites.[29] As also discussed in chapters 8 and 17, Cooper's hawks (*Accipiter*

THIS IS A NEST TREE OF A MISSISSIPPI KITE. THIS BIRD IS PROTECTED BY STATE AND
FEDERAL WILDLIFE LAWS AND THIS NEST IS BEING MONITORED BY RESEARCHERS FROM
THE TEXAS COOPERATIVE FISH AND WILDLIFE RESEARCH UNIT.

THE PRIMARILY INSECT DIET OF THIS SMALL FALCON-LIKE BIRD MAKES IT A BENEFICIAL
AND PLEASING WILDLIFE SPECIES TO HAVE AND WATCH IN OUR CITY. HOWEVER, IT WILL
ATTEMPT TO PROTECT ITS NESTLINGS WHEN HARASSED. PLEASE DO NOT DISTURB IT.

MISSISSIPPI KITES ARE
ONLY PRESENT IN TEXAS
DURING MAY–SEPTEMBER.
THEY MIGRATE TO SOUTH
AMERICA TO SPEND
THE WINTER MONTHS.

PLEASE ENJOY THE URBAN
WILDLIFE OF LUBBOX. IT IS A
COMMUNITY RESOURCE FOR
ALL TO ENJOY.

Figure 6.3. Signs alerting pedestrians to the presence of Mississippi kites and explaining why they may be aggressive can be important tools in gaining understanding and acceptance in urban areas.

cooperii),[33] red-shouldered hawks (*Buteo lineatus*),[34] and others have been documented swooping and diving at pedestrians near nesting areas. As raptors continue to colonize and exploit novel urban environments, some further conflict between raptors and humans is inevitable.

Although conflict solutions for one species are unlikely to be completely transferable to another, the well-documented efforts to eliminate or minimize aggressive behaviors of urban-nesting Mississippi kites may serve as a starting point for other species. Translocation of young and the use of effigies will be as time- and labor-intensive with other species as they were with Mississippi kites. Trimming nest trees or the removal of nests may be a viable management tool if raptors are nesting in areas where the public will not tolerate their presence. This technique, however, may simply move rather than minimize aggressive encounters. In the long-term, public education and awareness campaigns that focus on informing the public where aggressive raptors are nesting and why they are behaving aggressively are likely to be met with success.

Literature Cited

1. Parker, J. W., and J. C. Ogden. 1979. "The Recent History and Status of the Mississippi Kite." *American Birds* 33:119–29.
2. Kalla, P. I., and F. J. Alsop. 1983. "The Distribution, Habitat Preference, and Status of the Mississippi Kite in Tennessee." *American Birds* 37:146–49.

3. Sutton, G. M. 1939. "The Mississippi Kite in Spring." *Condor* 41:41–53.

4. Love, D., and F. L. Knopf. 1978. "The Utilization of Tree Plantings by Mississippi Kites in Oklahoma and Kansas." Proceedings of the 13th Annual Meeting of the Forestry Commission. *Great Plains Agricultural Council Publication 87*, Lincoln, NE.

5. Parker, J. W. 1996. "Urban Kites." In *Raptors in Human Landscapes: Adaptations to Built and Cultivated Environments*, edited by D. Bird, D. Varland, and J. J. Negro, 45–52. San Diego: Academic Press.

6. Evans, S. A. 1981. Ecology and Behavior of the Mississippi Kite (*Ictinia mississippiensis*) in Southern Illinois. MS thesis, Southern Illinois University.

7. Barber, J. D., E. P. Wiggers, and R. B. Renken. 1998. "Nest Site Characterization and Reproductive Success of Mississippi Kites in the Mississippi River Floodplains." *Journal of Wildlife Management* 62:1373–78.

8. Chiavacci, S. J., S. Schaefer, and J. C. Bednarz. 2009. "Nesting Ecology and Habitat Use of Swallow-Tailed and Mississippi Kites in the White River National Wildlife Refuge, Arkansas." Final Report to Arkansas Game and Fish Commission, Jonesboro, AR.

9. Bader, T. J., and J. C. Bednarz. 2010. "Home Range, Habitat Use, and Nest Site Characteristics of Mississippi Kites in the White River National Wildlife Refuge, Arkansas." *Wilson Journal of Ornithology* 122:706–15.

10. Parker, J. W. 1974. "The Breeding Biology of the Mississippi Kite in the Great Plains." PhD diss., University of Kansas.

11. Love, D, J. A. Grzybowski, and F. L. Knopf. 1985. "Influence of Various Land Uses on Windbreak Selection by Nesting Mississippi Kites." *Wilson Bulletin* 97:561–65.

12. Skipper, B. R. 2013. "Urban Ecology of Mississippi Kites." PhD diss., Texas Tech University.

13. Glinski, R. L., and R. D. Ohmart. 1983. "Breeding Ecology of the Mississippi Kite in Arizona." *Condor* 85:200–207.

14. Shaw, D. M. 1985. "The Breeding Biology of Urban-Nesting Mississippi Kites (*Ictinia mississippiensis*) in West Central Texas." MS thesis, Angelo State University.

15. Bolen, E. G., and D. L. Flores. 1989. "The Mississippi Kite in the Environmental History of the Southern Great Plains." *Prairie Naturalist* 21:65–74.

16. Welch, B. C., and C. W. Boal. 2015. "Prey Use and Provisioning Rates of Urban-Nesting Mississippi Kites in West Texas." *Journal of Raptor Research* 49:141–51.

17. Chiavacci, S. J., J. C. Bednarz, and T. J. Benson. 2014. "Does Flooding Influence the Types and Proportions of Prey Delivered to Nestling Mississippi Kites?" *Condor* 116:215–25.

18. Gennaro, A. L. 1986. "Breeding Biology of an Urban Population of Mississippi Kites in New Mexico." In *Proceedings of the Southwestern Raptor Management Symposium and Workshop*, edited by R. L. Glinski, B. G. Pendleton, M. B. Moss, M. N. LeFranc Jr., B. A. Millsap, and S. W. Hoffman, 188–190. Washington, DC: National Wildlife Federation.

19. Bolen, E. G., and D. L. Flores. 1993. *The Mississippi Kite*. Austin: University of Texas Press.

20. Parker, J. W. 1999. "Mississippi Kite (*Ictinia mississippiensis*)." In *The Birds of North America*, edited by P. G. Rodewald. Ithaca: Cornell Lab of Ornithology. Accessed March 3, 2017. https://birdsna.org/Species-Account/bna/species/miskit.

21. See chapter 4.

22. Welch, B. C. 2016. "The Breeding Ecology and Predicted Influence of Climate Change on Urban-Nesting Mississippi Kites." PhD diss., Texas Tech University.

23. St. Pierre, A. M. 2006. "The Breeding Ecology of Swallow-Tailed (*Elanoides forficatus*) and Mississippi Kites (*Ictinia mississippiensis*) in Southeastern Arkansas." MS thesis, Arkansas State University.

24. Stracey, C. M., and S. K. Robinson. 2012. "Does Nest Predation Shape Urban Birds Communities?" In *Urban Bird Ecology and Conservation*, edited by C. A. Lepczyk and P. S. Warren, 49–66. Berkeley: University of California Press.

25. Boal, C. W. 2008. "Hacked Nestling Mississippi Kite Survives 6 Years and Demonstrates Philopatry." *Bulletin of the Texas Ornithological Society* 41:66–68.

26. Skipper, B. R., and C. W. Boal. 2016. "Mate-Fidelity, Site-Fidelity, and Apparent Annual Survival of Urban Mississippi Kites." Working paper, Texas Cooperative Fish and Wildlife Research Unit, Texas Tech University.

27. Bader, T. J., and J. C. Bednarz. 2007. "An Extra-Pair Copulation by Mississippi Kites." *Journal of Raptor Research* 41:252–53.

28. See chapter 17.

29. See chapter 15.

30. Gennaro, A. L. 1986. "Extent and Control of Aggressive Behavior toward Humans by Mississippi Kites." In *Proceedings of the Southwestern Raptor Management Symposium and Workshop*, edited by R. L. Glinski, B. G. Pendleton, M. B. Moss, M. N. LeFranc Jr., B. A. Millsap, and S. W. Hoffman, 249–52. Washington, DC: National Wildlife Federation.

31. Andelt, W. F. 1994. "Mississippi Kites." In *Prevention and Control of Wildlife Damage*, edited by S. E. Hygnstrom, R. M. Timm, and G. E. Larson, E-75–77. Lincoln: University of Nebraska–Lincoln.

32. Good, H. B., and D. M. Johnson. 1976. "Experimental Tree Trimming to Control and Urban Winter Blackbird Roost." Paper 51, Bird Control Seminars Proceedings. Houston, TX.

33. Boal, C. W., and R. W. Mannan. 1999. "Comparative Breeding Ecology of Cooper's Hawk in Urban and Exurban Areas of Southeastern Arizona." *Journal of Wildlife Management* 63:77–84.

34. Dykstra, C. R., J. L. Hays, and S. T. Crocoll. 2008. "Red-Shouldered Hawk (*Buteo lineatus*)." In *The Birds of North America*, edited by P. G. Rodewald. Ithaca: Cornell Lab of Ornithology. Accessed August 17, 2017. https://birdsna.org/Species-Account/bna/species/reshaw.

CHAPTER 7

Cooper's Hawks:
The Bold Backyard Hunters

Robert N. Rosenfield, R. William Mannan, and Brian A. Millsap

WHAT A PLEASANT IRONY THAT a raptor traditionally regarded as a forest species and deemed to be headed for extinction in the last century due to loss of woodlands is now perhaps the most common backyard breeding raptor in (nonforested) cities throughout North America. Indeed, city residents commonly recount to us the dazzling flying speed of the "blue darter" in its pursuit of birds at feeders. Who would have guessed that city planners and the pervasive urban bird-feeding public conceivably (and unwittingly) contributed to the recovery of this red-eyed, blue-backed predator that is so boldly tolerant of the myriad of human activities in cities?

This intrepid darter, the Cooper's hawk (*Accipiter cooperii*), is a crow-sized raptor that typically preys on avian species such as doves (e.g., mourning doves [*Zenaida macroura*]) and small- to medium-sized songbirds (e.g., house sparrows [*Passer domesticus*] and American robins [*Turdus migratorius*]) that primarily forage on or near the ground; small, ground-dwelling mammals, especially the eastern chipmunk (*Tamias striatus*), are also important prey in some areas.[1,2,3] This hawk exhibits extreme reversed sexual size dimorphism compared to most raptor species, with females about 1.7 times heavier than males.[2,4] Its short, powerful, rounded wings and long tail allow for quick pursuit and marked maneuverability when chasing its agile bird prey. The Cooper's hawk is broadly distributed across temperate North America, ranging from both coasts north into southern Canada and south into Florida and northern Mexico.[2] Accounts from the 1890s and early 1900s suggest that this hawk was a common breeder throughout this range[5,6] and that it occasionally occurred in winter in southern Canada during the 1890s.[5] This territorial species almost exclusively selects trees for nesting in both exurban and urban habitats.[2,7,8,9]

Population Status

Cooper's hawk populations declined substantially during the mid-1900s, apparently due to multiple factors such as deforestation, persecution, eggshell thinning and aberrant adult courtship behavior due to ingestion of DDE (a primary breakdown product of DDT) present in songbird prey,[2] and increased mortality due to dieldrin exposure.[10] Perhaps the best evidence of these declines was a reduction in long-term autumnal migrant counts at Hawk Mountain in Pennsylvania,[11] concurrent with reduced reproductive success in some eastern states.[12] Although many states lacked conventional data to assess species' status metrics such as abundance/density, reproductive rate, and presence of contaminants in eggs or adults,[13,14] many states deemed it prudent to formally list the Cooper's hawk as extirpated, endangered, threatened, or as a species of special concern.[15,16] By the mid-1980s, Cooper's hawks were included on the imperiled or special concern lists of 15 of the 48 contiguous states.[17]

Cooper's hawk populations increased substantially after the 1972 ban on the use of DDT in the United States and their receiving protection under the Migratory Bird Treaty Act.[2] Tallies of autumnal migrating Cooper's hawks at various geographical watch and trapping sites reflected these increases.[11,18,19] Trend analyses from the North American Breeding Bird Survey (BBS) for 1966–2013 indicated a survey-wide increase of about 2 percent annually.[18]

Cooper's hawks seemingly reestablished breeding numbers across North America in their historical habitats of deciduous, coniferous, and mixed woodlands and forests and in sparsely wooded tracts in the grasslands of the northern Great Plains; they also were found nesting in pine plantations.[2,20] Urban and suburban development in Wisconsin and British Columbia (according to raptor researcher Frank Beebe) encroached into formerly natural Cooper's hawk habitat without loss of some breeding territories. Nesting Cooper's hawks also colonized urban environments where breeding habitat formerly did not exist (e.g., Tucson, Arizona, and Albuquerque, New Mexico).

Cooper's hawks predominantly nest in some conifer species compared to random samples of mixed tree species and disproportionately use conifer plantations compared to surrounding deciduous trees. These observations suggest a preference for nesting in conifers in some areas, including some urban settings (e.g., Douglas firs [*Pseudotsuga menziesii*] in British Columbia[21] and white pines [*Pinus strobus*] in Wisconsin).[22,23]

Some research has suggested that habituation to humans may jeopardize survival of Cooper's hawks[24] or that *Accipiter*, having evolved in shaded forest environments, may have low tolerances for higher temperatures and direct sunlight[25] or may not breed in highly fragmented forests.[26] Despite these speculations, since the 1970s, Cooper's hawks have become common, successful breeders in highly fragmented urban and suburban landscapes throughout North America, where human disturbance can be extreme and ambient temperatures are typically higher than in more shaded rural settings.[26,27,28,29,30,31] Perhaps surprisingly, some of the highest reported nesting densities and reproductive rates for the species occur in urban or suburban environs (color plate 2). Nest sites in these settings occur in virtually all environments, such as city parks, street boulevards, golf courses, cemeteries, and small forest fragments in industrial and commercial areas (figure 7.1). Cooper's hawks appear able to increase in numbers relatively quickly in urban settings. For example, the number of breeding attempts in a colonizing population nearly tripled (with a concomitant increase in reproductive rate) in just twelve years in Milwaukee, Wisconsin.[8]

As long ago as the late 1890s, there were anecdotal reports that Cooper's hawks had become increasingly common during winter in "larger parks" of cities where house sparrows served as prey.[5] Cooper's hawks are currently resident year-round in many northern US cities and in Victoria, British Columbia, where winter migration likely had been typical previously.[2,32,33] These reports suggest marked flexibility in habitat use and, perhaps, migration behavior of a raptor formerly regarded as secretive and in immediate demographic peril. Indeed, the Cooper's hawk is likely the most common raptor seen at bird feeders or nesting in residential settings throughout North America.[34] The popular press has chronicled some of this species' urban ecology;[35,36] one can readily find YouTube videos of Cooper's hawks hunting or drinking at birdbaths in backyards. Unsurprisingly, the species is, to our knowledge, no longer formally listed as a species of conservation concern by any US state or Canadian province.

Figure 7.1. Cooper's hawk nest in urban Milwaukee, Wisconsin. Almost nonstop traffic and regular pedestrians do not deter Cooper's hawks from nesting in the median of this highly traveled road. Photo by William Stout.

Urban areas are relatively new environments for most breeding populations of Cooper's hawks and other raptors, and the fundamental ecology of these populations is generally poorly studied, especially on a long-term basis.[24,37,38] There are, however, notable exceptions, such as the multidecadal, cross-generational investigations of the breeding biology of Cooper's hawks in the cities of Tucson, Arizona;[39,40] Victoria, British Columbia;[27] and Stevens Point and Milwaukee, Wisconsin.[8,41,42] These studies provide strong bases for assessing how this species is adapting or has adapted to a human-dominated environ that is ever-encroaching into most terrestrial ecosystems in North America. We highlight these and other studies that demonstrate the species' use of urban environments and the ecological research that helps elucidate selective pressures that may differ from those influencing exurban Cooper's hawks.[43] We emphasize that urban landscapes vary in size, land-cover heterogeneity, prey populations, and other factors that potentially affect the ecology of raptors; such high variability of urban environs may cause variation in the ecology of the Cooper's hawk across North America.[27,44,45]

Urban Behavioral Ecology

When animals colonize a new environment such as a city, the novel landscape often influences the pioneering individuals,[46] modifying their behavior compared to exurban birds.[47] It is possible that several previously undocumented behaviors of the nesting biology of Cooper's hawks—such as polygyny, wherein one male successfully breeds simultaneously with two females, each one at a different nest;[48] nest helpers, in which a second male tends and/or aggressively defends a nest;[28] and hunting of roosting prey by illumination of urban lighting and the moon[32]— may result from (or be a release from) selective pressures in novel, city environments in which these behaviors were recorded.[46] Most raptors are generally monogamous, with occasional extra-pair paternity. Counter to this pattern, the highest rate in raptors of extra-pair paternity (34 percent of 44 study nests) was found in Cooper's hawks breeding in Milwaukee, Wisconsin.[49] The high rate of extra-pair paternity in Milwaukee may be related to the synergistic effects of the strong association between courtship feeding and copulations in Cooper's hawk and a high nesting density in a food-rich urban setting.[49] Specifically, females may be maximizing energy intake by trading copulations for food from territorial and nonterritorial males on both their own and neighboring territories during the pre-incubation period.[49,50] Alternatively, it is possible that some of the above behaviors also occur in exurban settings but are simply more readily observed in the relatively open, sparsely wooded environments of urban settings. Indeed, preliminary analyses suggest comparably high extrapaternity rates in Cooper's hawks breeding in urban and exurban environs of central Wisconsin, according to recent studies done in collaboration with S. A. Sonsthagen and M. G. Hardin. However, breeding adult Cooper's hawks vocalized less during feeding in urban versus rural environments in and around Tucson, Arizona.[51] This may reflect differences in food stress experienced by urban birds, with females being less likely to vocalize at an urban nest because of lower food stress compared to rural sites. More comparative studies of urban and exurban ecology are needed to better understand the influence of city environments on life histories of raptors. Further, very few studies directly address the underlying causes of behavioral differences between urban and non-urban animals, and thus we need detailed studies in which the effects of selection, plasticity, and sorting can be disentangled. The study of raptors from more natural, biodiverse regions that are still in the early stages of becoming urban-dwellers[52] would be especially useful. As suggested in chapter 2, species or individuals that can adjust their behaviors to the new selection pressures presented by cities should have greater success in urban environments.[53] Indeed, a rare polygynous Cooper's hawk contributed a disproportionate number of recruits to an urban population in Grand Forks, North Dakota.[54]

Urban Habitat Suitability and Selected Population Demographics

Suitability of habitat is generally considered to be the primary factor influencing the viability of breeding populations of birds,[13,55] and thus it is important to measure via demographic data, such as adult survival rates, breeding densities, and reproductive rates, the suitability of urban habitats for Cooper's hawks. It is possible that anthropogenic factors in urban environs may render some cities less suitable than exurban sites.[24,26,56] For example, as discussed in chapter 14, mortality rates may be higher for urban birds due to collisions with anthropogenic obstacles (e.g., windows, electrical utility wires, cars).[31,32,57] Alternatively, the greater abundance of food resources in urban (compared to rural) environments[30] could result in enhanced survival;[58] this food-survivorship theme may be particularly pertinent for raptors because starvation is an important cause of mortality for birds of prey.[14] However, different mortality rates in cities versus exurban areas have not been conclusively documented in Cooper's hawk populations. There was no significant difference in annual survivorship of adult males breeding in urban (84 percent) and rural (79 percent) areas of Wisconsin. Further, there was no significant temporal variation across 26 years in overall annual survivorship rate (approximately 81 percent) of nesting males in Wisconsin;[59] breeding females had approximately 75 percent annual survivorship in the same study areas.[14] These survival rates are similar to the rates of 80 percent reported for both nesting male and female Cooper's hawks in Tucson, Arizona,[60] and to a 75 percent survival rate of radio-tagged adult male and female Cooper's hawks over a 110-day interval during winter in rural and urban Indiana (rates not separated by habitat).[32] These rates are also similar to the annual survivorship rates of approximately 80 percent for both breeding males and females in rural Florida.[3] These similar and relatively high survivorship metrics across a large geographical scale of North America are consistent with apparently healthy *Accipiter* populations,[60] and perhaps the similarity of survival rates suggests that habitat type does not appreciably affect the mortality rate of adults. However, we note that researchers compared urban to exurban mortality rates of Cooper's hawks in only one study,[59] and we suggest the comparative influence of these two habitats on survivorship of breeding adults warrants additional research.

Compounding the issue is the fact that most studies were unable to document many of the causes of mortality,[59] further accentuating our inability to understand the correlates of mortality in different environments. However, studies in Indiana[32] and in California[61] suggested that predation on Cooper's hawks was more common in rural compared to urban sites, although sample sizes were small (see also the study in Tucson, Arizona).[62] Conversely, predation rates did not differ between urban and exurban Cooper's hawk nests in southeastern Wisconsin.[28]

Cooper's hawk nesting densities vary markedly, with reported densities of 1 nest per 671–2,326 ha in the western states[39,63] and 1 nest per 272–5,000 ha in midwestern and eastern states.[3,8,15] The highest reported nesting densities have been found in cities (e.g., 1 nest/272 ha in Stevens Point, Wisconsin [with a human population of about 38,000 people; figure 7.2]; 1 nest/437 ha in Tucson, Arizona [approximately 900,000 people]; and 1 nest/330 ha in Milwaukee, Wisconsin [approximately 1,000,000 people]). Closest neighboring pairs may be, on average, only about 800 m apart at higher densities for this territorial species, and in Albuquerque, New Mexico, and Victoria, British Columbia, pairs nested successfully within 160 m and 300 m of each other in studies conducted by the authors and raptor researcher Andrew Stewart. In contrast, nearest nesting pairs in exurban areas are generally no closer than about 2.6 km apart. Thus Cooper's hawks may attain higher densities in urban compared to exurban areas.[64] Although several hypotheses have been offered to explain this disparity, including reduced predation and availability of water in cities,[64] it is generally believed that greater food availability in urban environments is the causative agent.[29,58] The number of individuals and overall biomass of avian prey for this bird-eating hawk are typically higher in urban environments compared to exurban environments.[29] Indeed, with the exception of one study,[61] telemetry studies suggest that urban breeding and wintering Cooper's hawks do not range as far as their rural counterparts to obtain adequate food.[3,40,65,66] Interestingly, introduced species, such as house sparrows and European starlings (*Sturnus vulgaris*), both of which cause damage to native ecosystems and species,[29,67] can provide an important food source for native breeding Cooper's hawks.[44]

Urban Cooper's hawks tend to have higher reproductive rates than those in more typical rural settings, although this is not consistent across all studies. The average brood size of Cooper's hawks ranges from 1.5 to 4.0 young per successful nest (i.e., a nest with fledged young or with young at least 70 percent of fledging age).[2,25,61,68,69] Some of the highest reproductive rates, with averages ranging from 3.6 to 4.0 young per successful nest, are from urban settings[27,39,41] and exceed the estimates ranging from 2.8 to 3.0 in exurban studies.[3,25,70] We caution, however, that lower productivity indices do not necessarily indicate

Figure 7.2. Distribution of Cooper's hawks on the urban 3,540 ha study plot in Stevens Point, Wisconsin, 1993. The nesting density of 272 ha/pair is the highest reported nesting density for this species. Note the close spacing of about 0.8 km between nests in the southernmost part of the study site. Comparably high nesting densities were found in Tucson, Arizona; Milwaukee, Wisconsin; and Victoria, British Columbia.

lower population viability and/or lower habitat quality. Indeed, lower reproductive rates in a western North Dakota population were apparently not related to increased hatchling mortality rates, limited food, or reduced nest-site availability. Researchers hypothesized that the lower reproductive rates may have been the result of a trade-off between the energetic demands of a long migration and the phenological constraints of a short breeding season, which limited the ability of breeding birds to accumulate energetic reserves for egg production.[27] In fact, this population was viable and increasing. These findings suggest that local ecological and life history factors vary widely among populations of Cooper's hawks, even for those that are stable or growing.

Comparative studies of the reproductive rates of urban versus exurban Cooper's hawk populations are few, but the results of three such studies varied. Two reported higher than average numbers of fledged young in urban versus exurban nests: 3.1 versus 2.8 in Arizona[68] and 3.7 versus 1.5 in California;[61] this latter study was limited by very small samples sizes. These valuable paired-sample studies suggest that some urban habitats may be of higher quality than nearby exurban habitats. In contrast, there were no significant differences in average clutch counts (approximately 4.3 eggs) and average brood sizes (approximately 3.6 young) between nests in urban and exurban habitats across 36 years in Wisconsin.[23,41,42] Nesting phenology did not differ between urban and exurban nests in Wisconsin,[14,42] which contrasted with a study in Arizona in which urban birds nested 16 days earlier than exurban ones.[68] As also noted in chapters 2 and 4, urban landscapes vary greatly in size, heterogeneity, prey populations, and other ecological factors (including surrounding environments) that potentially affect reproductive success of raptors. Such high variability of urban environs may explain the variation in the ecology of the Cooper's hawk across North America.[27,42]

Another reproductive metric with no consistent urban-rural difference was nest success, the proportion of nests with eggs (i.e., incubating adults were observed) that produced nestlings at least two weeks of age. Nest success ranged from 47 percent to 91 percent, with most around 70 percent.[2,25,27,68,69] There was no general pattern suggesting that urban Cooper's hawks have either higher or lower nest success than exurban ones. There is, however, some evidence that nests placed closer to the ground may have greater predation risk than nests placed higher above the ground.[27,64] It is difficult to generalize about causes of nest failures of Cooper's hawks, as most nests fail for unknown reasons;[27] however, predation at nests by red-tailed hawks (*Buteo jamaicensis*), great horned owls (*Bubo virginianus*), and raccoons (*Procyon lotor*) seems important and widespread.[2,8] We also note that Cooper's hawks are still shot, perhaps more frequently in some urban environs where hawks are less wary of humans, but the extent of this illegal activity is inherently difficult to measure.

Ecological Traps

An ecological trap is a concept related to habitat quality and is potentially a concern for conservation of species occupying urban areas.[71,72] In an ecological trap, environmental cues used by animals to select a place to live suggest it is quality habitat, but in reality, circumstances such as novel agents of mortality (e.g., exotic diseases, electrocution, window strikes) reduce reproduction or survival to the point that the population cannot sustain itself without immigration from elsewhere.[71] Anthropogenic activities may cause most traps,[72] and it therefore seems plausible that animals occupying or possibly lured to urbanized environments may be especially susceptible to ecological traps. To our knowledge, no study has demonstrated the existence of an urban (or exurban) environment as a trap for Cooper's hawks in North America.

High mortality (approximately 40 percent) of nestlings due to trichomoniasis, a disease that that is caused by a parasitic protozoan (*Trichomonas gallinae*) in doves and other birds eaten by Cooper's hawks,[51] prompted initial speculation that Tucson, Arizona, was an ecological trap for the species.[68] Despite the prevalence of the urban-related disease, further analyses based on high productivity and high rates of survival, particularly of hawks in their first year of life if they survived the disease,[73] demonstrated that the population was growing rapidly and that Tucson was not a trap for Cooper's hawks.[60] Elsewhere across northern North America, urban and exurban nestlings and breeding adult Cooper's hawks had low rates of infection (<3 percent of 257 birds sampled) and no nestling deaths attributable to trichomoniasis.[74] In addition, the percentage of nestling Cooper's hawks in Tucson with trichomonads decreased from 85 percent in the mid-1990s to 16 percent in 2010 and 2011.[75]

Although not specifically urban in structure, pine (*Pinus* spp.) plantations are a human-altered habitat occurring in both urban and exurban sites. Although pine plantations are thought to be inferior in habitat quality, Cooper's hawks nesting there contributed recruits to subsequent generations in proportion to their production of young in Wisconsin.[23] Nesting density and productivity indices in urban and exurban plantations were among the highest reported for Cooper's hawks, suggesting that pine plantations were high-quality breeding habitat and that they were not ecological traps for the species in Wisconsin.[23]

Conservation and Management

The combination of continental distribution, behavioral plasticity, marked habitat flexibility and relatively high adult survivorship, nesting densities, and reproductive success in urban and exurban environments appears to have been propitious for the Cooper's hawk in North America. However, the behavioral and population responses to urbanization are understudied in raptors, including the Cooper's hawk.

Dates of spring migration of Cooper's hawks and other raptors in the Great Lakes region have advanced, associated with recent climate change,[76] prompting researchers to express concern for the viability of raptor populations.[76] Earlier migration may lead to earlier breeding, which could result in mismatches between reproductive timing and food availability. Such decoupling may be more severe for species at higher trophic levels, such as raptors.[77] Six consecutive generations of Wisconsin Cooper's hawks exhibited advanced timing of egg laying in urban and exurban habitats during 1980–2015; however, their high reproductive rate was not affected.[77]

We suggest that the widespread presence of the Cooper's hawk in both urban and exurban habitats makes it an excellent study species to investigate the poorly documented mechanisms, including evolutionary forces, that influence the manner in which a raptor survives and reproduces in an urbanized world.[4,78] In addition, its readily observed urban presence could perhaps facilitate its potential role as an ambassador to the public to enhance conservation efforts that are in part dependent on public education in science and biology.[79]

Acknowledgments

The Department of Biology, the Personnel Development Committee, and the Letters and Science Enhancement Fund at the University of Wisconsin at Stevens Point provided funding and sabbatical support for R. N. Rosenfield.

Literature Cited

1. Bielefeldt, J., R. N. Rosenfield, and J. M. Papp. 1992. "Unfounded Assumptions about Diet of the Cooper's Hawk." *Condor* 94:427–36.
2. Rosenfield, R. N., and J. Bielefeldt. 1993. "Cooper's Hawk (*Accipiter cooperii*)." In *The Birds of North America*, no. 75, edited by A. Poole and F. Gill. Philadelphia: The Academy of Natural Sciences; Washington, DC: American Ornithologists' Union.
3. Millsap, B. A., T. F. Breen, and L. M. Phillips. 2013. "Ecology of the Cooper's Hawk in North Florida." *North American Fauna* 78:1–58.
4. Rosenfield, R. N., L. J. Rosenfield, J. Bielefeldt, R. K. Murphy, A. C. Stewart, W. E. Stout, T. G. Driscoll, and M. A. Bozek. 2010. "Comparative Morphology of Northern Populations of Breeding Cooper's Hawks." *Condor* 112:347–55.
5. Fisher, A. K. 1893. "The Hawks and Owls of the United States in Their Relation to Agriculture." US Dept. of Agriculture, Division of Ornithology and Mammalogy, Bulletin No. 3, Washington, DC: GPO.

6. Bent, A. C. 1937. "Part 1." *Life Histories of North American Birds of Prey.* New York: Dover Publications, 112–125.

7. Boal, C. W., and R. W. Mannan. 1998. "Nest-Site Selection by Cooper's Hawks in an Urban Environment." *Journal of Wildlife Management* 62:864–71.

8. Stout, W. E., and R. N. Rosenfield. 2010. "Colonization, Growth, and Density of a Pioneer Cooper's Hawk Population in a Large Metropolitan Environment." *Journal of Raptor Research* 44:255–67.

9. Bird, D. M., D. E. Varland, and J. J. Negro, eds. 1996. *Raptors in Human Landscapes: Adaptations to Built and Cultivated Environments.* San Diego: Academic Press.

10. Nisbet, I. C. T. 1988. "The Relative Importance of DDE and Dieldrin in the Decline of Peregrine Falcon Populations." In *Peregrine Falcon Populations: Their Management and Recovery*, edited by T. J. Cade, J. H. Enderson, C. G. Thelander, and C. M. White, 351–75. Boise, ID: The Peregrine Fund.

11. Bednarz, J., D. Klem, L. J. Goodrich, and S. E. Senner. 1990. "Migration Counts of Raptors at Hawk Mountain, Pennsylvania, as Indicators of Population Trends, 1934–1986." *Auk* 107:96–109.

12. Henny, C. J., and H. M. Wight. 1972. "Population Ecology and Environmental Pollution: Red-Tailed and Cooper's Hawks." *Population Ecology of Migratory Birds: A Symposium*, 229–50. US Fish and Wildlife Service Report 2, Washington, DC.

13. Andersen, D. E., S. DeStefano, M. I. Goldstein, K. Titus, D. C. Crocker-Bedford, J. J. Keane, R. G. Anthony, and R. N. Rosenfield. 2004. "The Status of Northern Goshawks in the Western United States." *Technical Review* 39 (3):192–209, Bethesda: Wildlife Society.

14. Rosenfield, R. N., J. Bielefeldt, T. G. Haynes, M. G. Hardin, F. J. Glassen, and T. L. Booms. 2016. "Body Mass of Female Cooper's Hawks Is Unrelated to Longevity and Breeding Dispersal: Implications for the Study of Breeding Dispersal." *Journal of Raptor Research* 50:305–12.

15. Rosenfield, R. N., J. Bielefeldt, R. K. Anderson, and J. M. Papp. 1991. "Status Reports: Accipiters." In *Proceedings of the Midwest Raptor Management Symposium and Workshop*, edited by B. G. Pendleton and D. L. Krahe, 42–49. Washington, DC: National Wildlife Federation.

16. Rosenfield, R. N., M. G. Hardin, J. Bielefeldt, and R. K. Anderson. 2016. "Status of the Cooper's Hawk in Wisconsin, Wisconsin Endangered Resources Report Number 8: With Selective Retrospective and Interpretation." *Passenger Pigeon* 78:191–200.

17. LeFranc, M. N., and B. A. Millsap. 1984. "A Summary of State and Federal Agency Raptor Management Programs." *Wildlife Society Bulletin* 12:272–82.

18. Sauer, J. R., J. E. Hines, J. E. Fallon, K. L. Pardieck, D. J. Ziolkowski, and W. A. Link. 2014. "The North American Breeding Bird Survey, Results and Analysis 1966–2012." Version 02.19.2014. Accessed September 17, 2014. http://www.mbr-pwrc.usgs.gov/bbstr2012/tr03330.htm.

19. Mueller, H. C., D. D. Berger, G. Allez, N. S. Mueller, W. R. Robichaud, and J. L. Kaspar. 2001. "Migrating Raptors and Vultures at Cedar Grove, Wisconsin, 1936–1999: An Index to Population Changes." In *Hawkwatching in the Americas*, edited by K. L. Bildstein and D. Klem, 1–22. North Wales, ME: Hawk Migration Association of North America.

20. Murphy, R. K. 1993. "History, Nesting Biology, and Predation Ecology of Raptors in the Missouri Coteau of Northwestern North Dakota." PhD diss., Montana State University.

21. Campbell, R. W., N. K. Dawe, I. McTaggert-Cowan, J. M. Cooper, G. W. Kaiser, and M. C. E. McNall. 1990. *The Birds of British Columbia, Vol. 2: Nonpasserines*. Vancouver, BC: Mitchell Press.

22. Trexel, D. R., R. N. Rosenfield, J. Bielefeldt, and E. G. Jacobs. 1999. "Comparative Nest Site Habitats in Sharp-Shinned and Cooper's Hawks in Wisconsin." *Wilson Bulletin* 111:7–14.

23. Rosenfield, R. N., J. Bielefeldt, S. A. Sonsthagen, and T. L. Booms. 2000. "Comparable Reproductive Success at Conifer and Non-Plantation Nest Sites for Cooper's Hawks in Wisconsin." *Wilson Bulletin* 112:417–21.

24. Snyder, H. A., and N. F. R. Snyder. 1974. "Increased Mortality of Cooper's Hawks Accustomed to Man." *Condor* 76:215–16.

25. Reynolds, R. T., E. C. Meslow, and H. M. Wight. 1982. "Nesting Habitat of Coexisting Accipiter in Oregon." *Journal of Wildlife Management* 46:124–38.

26. Bosakowski, T., R. Speiser, D. G. Smith, and L. J. Niles. 1993. "Loss of Cooper's Hawk Nesting Habitat to Suburban Development: Inadequate Protection for a State-Endangered Species." *Journal of Raptor Research* 27:36–40.

27. Rosenfield, R. N., J. Bielefeldt, L. J. Rosenfield, A. C. Stewart, M. P. Nenneman, R. K. Murphy, and M. A. Bozek. 2007b. "Variation in Reproductive Indices in Three Populations of Cooper's Hawks." *Wilson Journal of Ornithology* 119:181–88.

28. Stout, W. E., R. N. Rosenfield, W. G. Holton, and J. Bielefeldt. 2007. "Nesting Biology of Urban Cooper's Hawks in Milwaukee, Wisconsin." *Journal of Wildlife Management* 71:366–75.

29. Chace, J. F., and J. J. Walsh. 2006. "Urban Effects on Native Avifauna: A Review." *Landscape and Urban Planning* 74:46–69.

30. Marzluff, J. M., F. R. Gehlbach, and D. A. Manuwal. 1998. "Urban Environments: Influences on Avifauna and Challenges for the Avian Conservationist." In *Avian Conservation: Research and Management*, edited by J. M. Marzluff and R. Salabanks, 283–99. Washington, DC: Island Press.

31. Grimm, N. B., S. H. Faeth, N. E. Golubiewski, C. L. Redman, J. Wu, X. Bai, and J. M. Briggs. 2008. "Global Change and the Ecology of Cities." *Science* 319:756–60.

32. Roth, T. C., S. L. Lima, and W. E. Vetter. 2005. "Survival and Causes of Mortality in Wintering Sharp-Shinned Hawks and Cooper's Hawks." *Wilson Bulletin* 117:237–44.

33. Meehan, T. D., R. N. Rosenfield, V. N. Atudorei, J. Bielefeldt, L. J. Rosenfield, A. C. Stewart, W. E. Stout, and M. A. Bozek. 2003. "Variation in Hydrogen Stable-Isotope Ratios between Adult and Nestling Cooper's Hawks." *Condor* 105:567–72.

34. Dunn, E. H., and D. L. Tessaglia. 1994. "Predation of Birds at Feeders in Winter." *Journal of Field Ornithology* 65:8–16.

35. Harrison, G. H. 2003. "Feeder Hawk; Cooper's Hawks Seek Out Backyard Bird Feeders, but Not for the Seed." *Birder's World*, October 2013, 36–39.

36. Bird, D. M. 2014. "Watching Bird Behavior: Introducing New Species." *Bird Watcher's Digest*, January/February 2014, 124–31.

37. Love, O. P., and D. M. Bird. 2000. "Raptors in Urban Landscapes: A Review and Future Concerns." In *Raptors at Risk: Proceedings of the V World Conference on Birds of Prey and Owls*, edited by R. D. Chancellor and B.-U. Meburg, 425–34. Berlin, Germany: World Working Group on Birds of Prey and Owls.

38. Rutz, C. 2008. "The Establishment of an Urban Bird Population." *Journal of Animal Ecology* 77:1008–19.

39. Mannan, R. W., R. N. Mannan, C. A. Schmidt, and W. A. Estes-Zumph. 2007. "Influence of Natal Experience on Nest-Site Selection by Urban-Nesting Cooper's Hawks." *Journal of Wildlife Management* 71:64–68.

40. Boggie, M. A., and R. W. Mannan. 2014. "Examining Seasonal Patterns of Space Use to Gauge How an Accipiter Responds to Urbanization." *Landscape and Urban Planning* 124:34–42.

41. Rosenfield, R. N., J. Bielefeldt, J. L. Affeldt, and D. J. Beckmann. 1995. "Nesting Density, Nest Area Reoccupancy, and Monitoring Implications for Cooper's Hawks in Wisconsin." *Journal of Raptor Research* 29:1–4.

42. Rosenfield, R. N., W. E. Stout, M. D. Giovanni, N. H. Levine, J. A. Cava, M. G. Hardin, and T. G. Haynes. 2015b. "Does Breeding Population Trajectory and Age of Nesting Females Influence Disparate Nestling Sex Ratios in Two Population of Cooper's Hawks?" *Ecology and Evolution* 5:4037–48.

43. See chapter 18.

44. Cava, J. A., A. C. Stewart, and R. N. Rosenfield. 2012. "Introduced Species Dominate the Diet of Breeding Urban Cooper's Hawks in British Columbia." *Wilson Journal of Ornithology* 124:775–82.

45. Sumasgutner, P., E. Nemeth, G. Tebb, H. W. Krenn, and A. Gamauf. 2014. "Hard Times in the City—Attractive Nest Sites but Insufficient Food Supply Lead to Low Reproduction Rates in a Bird of Prey." *Frontiers in Zoology* 11:48.

46. Price, T. D., P. J. Yeh, and B. Harr. 2008. "Phenotypic Plasticity and the Evolution of a Socially Selected Trait following Colonization of a Novel Environment." *American Naturalist* 172:S49–62.

47. Ditchkoff, S. S., S. T. Saalfeld, and C. J. Gibson. 2006. "Animal Behavior in Urban Ecosystems: Modifications Due to Human-Induced Stress." *Urban Ecosystems* 9:5–12.

48. Rosenfield, R. N., T. G. Driscoll, R. P. Franckowiak, L. J. Rosenfield, B. L. Sloss, and M. A. Bozek. 2007a. "Genetic Analysis Confirms First Record of Polygyny in Cooper's Hawks." *Journal of Raptor Research* 41:230–34.

49. Rosenfield, R. N., S. A. Sonsthagen, W. E. Stout, and S. L. Talbot. 2015a. "High Frequency of Extra-Pair Paternity in an Urban Population of Cooper's Hawks." *Journal of Field Ornithology* 86:144–52.

50. Lien, L. A., B. A. Millsap, K. Madden, and G. W. Roemer. 2015. "Male Brood Provisioning Rates Provide Evidence for Inter-Age Competition for Mates in Female Cooper's Hawks *Accipiter cooperii*." *Ibis* 157:860–70.

51. Estes, W. A., and R. W. Mannan. 2003. "Feeding Behavior of Cooper's Hawks at Urban and Rural Nests in Southeastern Arizona." *Condor* 105:107–16.

52. Sol, D., O. Lapiedra, and C. González-Lagos. 2013. "Behavioral Adjustments for Life in the City." *Animal Behaviour* 85:1101–12.

53. Lowry, H., A. Lill, and B. B. M. Wong. 2012. "Behavioral Responses of Wildlife to Urban Environments." *Biological Reviews* 88:537–49.

54. Driscoll, T. G., and R. N. Rosenfield. 2015. "Polygyny Leads to Disproportionate Recruitment in Urban Cooper's Hawks (*Accipiter cooperii*)." *Journal of Raptor Research* 49:344–46.

55. Cody, M. L. 1985. "An Introduction to Habitat Selection in Birds." In *Habitat Selection in Birds*, edited by M. L. Cody, 3–56. San Diego: Academic Press.

56. Mumme, R. L., S. J. Schoech, G. E. Woolfenden, and J. W. Fitzpatrick. 1999. "Life and Death in the Fast Lane: Demographic Consequences of Road Mortality in the Florida Scrub-Jay." *Conservation Biology* 14:501–12.

57. Sweeney, S. J., P. T. Redig, and H. B. Tordoff. 1997. "Morbidity, Survival and Productivity of Rehabilitated Peregrine Falcons in Upper Midwestern U.S." *Journal of Raptor Research* 31:347–52.

58. Chamberlain, D. E., A. R. Cannon, M. P. Toms, D. I. Leech, B. J. Hatchwell, and K. J. Gaston. 2009. "Avian Productivity in Urban Landscapes: A Review and Meta-Analysis." *Ibis* 151:1–18.

59. Rosenfield, R. N., J. Bielefeldt, L. J. Rosenfield, T. L. Booms, and M. A. Bozek. 2009. "Survival Rates and Lifetime Reproduction of Breeding Male Cooper's Hawks in Wisconsin, 1980–2005." *Wilson Journal of Ornithology* 121:610–17.

60. Mannan, R. W., R. J. Steidl, and C. W. Boal. 2008. "Identifying Habitat Sinks: A Case Study of Cooper's Hawks in an Urban Environment." *Urban Ecology* 11:141–48.

61. Chiang, S. N., P. H. Bloom, A. M. Bartuszevige, and S. E. Thomas. 2012. "Home Range and Habitat Use of Cooper's Hawks in Urban and Natural Areas." In *Urban Bird Ecology and Conservation*, edited by C. A. Lepczyk and P. S. Warren. Studies in Avian Biology, no. 45. Berkeley: University of California Press, 1–16. http://www.ucpress.edu/go/sab.

62. Boal, C. W. 1997. "The Urban Environment as an Ecological Trap for Cooper's Hawks." PhD diss., University of Arizona.

63. Reynolds, R. T. 1989. "Status Reports: Accipiters." In *Proceedings of the Midwest Raptor Management Symposium and Workshop*, edited by B. G. Pendleton, C. E. Ruibal, D. L. Krahe, K. Steenhof, M. N. Kochert, and M. L. LeFranc, 92–101. Washington, DC: National Wildlife Federation.

64. Boal, C. W., and R. W. Mannan. 1998. "Nest-Site Selection by Cooper's Hawks in an Urban Environment." *Journal of Wildlife Management* 62:864–71.

65. Mannan, R. W., and C. W. Boal. 2000. "Home Range Characteristics of Male Cooper's Hawks in an Urban Environment." *Wilson Bulletin* 112:21–27.

66. Roth, T. C., II, W. E. Vetter, and S. L. Lima. 2008. "Spatial Ecology of Wintering *Accipiter* Hawks: Home Range, Habitat Use, and the Influence of Bird Feeders." *Condor* 110:260–68.

67. Mack, R. N., D. Simberloff, W. M. Lonsdale, H. Evans, M. Clout, and F. Bazzaz. 2000. "Biotic Invasions: Causes, Epidemiology, Global Consequences and Control." *Issues in Ecology* 5:1–22.

68. Boal, C. W., and R. W. Mannan. 1999. "Comparative Breeding Ecology of Cooper's Hawks in Urban and Exurban Areas of Southeastern Arizona." *Journal of Wildlife Management* 63:77–84.

69. Nenneman, M. P., R. K. Murphy, and T. A. Grant. 2002. "Cooper's Hawks, *Accipiter cooperii*, Nest at High Densities in the Northern Great Plains." *Canadian Field-Naturalist* 116:580–84.

70. Conrads, D. J. 1997. "The Nesting Ecology of the Cooper's Hawk in Iowa." *Iowa Bird Life* 67:33–41.

71. Pulliam, H. R. 1988. "Sources, Sinks, and Population Regulation." *American Naturalist* 132:652–61.

72. Robertson, B. A., and R. L. Hutto. 2006. "A Framework for Understanding Ecological Traps and an Evaluation of Existing Evidence." *Ecology* 87:1075–85.

73. Mannan, R. W., W. A. Estes, and W. J. Matter. 2004. "Movements and Survival of Fledgling Cooper's Hawks in an Urban Environment." *Journal of Raptor Research* 38:26–34.

74. Rosenfield, R. N., S. J. Taft, W. E. Stout, T. G. Driscoll, D. L. Evans, and M. A. Bozek. 2009. "Low Prevalence of *Trichomonas gallinae* in Urban and Migratory Cooper's Hawks in North Central North America." *Wilson Journal of Ornithology* 121:641–44.

75. Urban, E. H., and R. W. Mannan. 2014. "The Potential Role of Oral pH in the Persistence of *Trichomoniasis gallinae* in Cooper's Hawks (*Accipiter cooperii*)." *Journal of Wildlife Diseases* 50:50–55.

76. Sullivan, A. R., D. J. Flaspohler, R. E. Froese, and D. Ford. 2016. "Climatic Variability and the Timing of Spring Raptor Migration in Eastern North America." *Journal of Avian Biology* 47:208–18.

77. Rosenfield, R. N., M. G. Hardin, J. Bielefeldt, and E. R. Keyel. 2016. "Are Life History Events of a Northern Breeding Population of Cooper's Hawks Influenced by Changing Climate?" *Ecology and Evolution* 6:399–408.

78. Sonsthagen, S. A., R. N. Rosenfield, J. Bielefeldt, R. K. Murphy, A. C. Stewart, W. E. Stout, T. G. Driscoll, M. A. Bozek, B. L. Sloss, and S. L. Talbot. 2012. "Genetic and Morphological Divergence among Cooper's Hawk (*Accipiter cooperii*) Populations Breeding in North-Central and Western North America." *Auk* 129:427–37.

79. Stracey, C. M. 2011. "Resolving the Urban Nest Predator Paradox: The Role of Alternative Foods for Nest Predators." *Biological Conservation* 144:1545–52.

CHAPTER 8

Red-Shouldered Hawks: Adaptable Denizens of the Suburbs

Cheryl R. Dykstra, Peter H. Bloom, and Michael D. McCrary

WITH ITS BEAUTIFUL BLACK-AND-WHITE WINGS and tail contrasting with a richly rusty body and its insistent "kee-kee-kee" calls, the red-shouldered hawk (*Buteo lineatus*) is a welcome neighbor for many suburban residents—a little bit of wilderness that inserts itself into everyday life. That it eats mainly rodents, snakes, and other species perceived as "pests" endears it to some suburbanites, while its young entertain others with their antics in the nest and their first, clumsy attempts at flight. But as with all complex relationships, the association between humans and red-shouldered hawks is not uniformly benign; hawks defending their nests may injure humans on the ground, and human activities, such as the use of rodenticides and continued development, may threaten red-shouldered hawk success in cities and suburbs.

In some ways, the red-shouldered hawk is an ideal candidate for inhabiting urban areas (a term we use broadly in this chapter to include the full gradient of urban to suburban environments). Successful urban species are often generalists with small area requirements,[1,2] and some behavioral flexibility.[2,3,4,5] The red-shouldered hawk is a medium-sized *Buteo* (with a mean wing chord of 282–345 mm and a mass of 486–774 g)[6] and is smaller than the red-tailed hawk (*Buteo jamaicensis*) and Swainson's hawk (*Buteo swainsoni*), both of which inhabit urban areas in some situations. It is a dietary generalist, capturing and consuming a variety of prey, including rodents, reptiles, amphibians, birds, fish, and invertebrates.[6] Red-shoulders nest in many tree species[6] and have relatively small home ranges, from 100 to 200 ha in natural habitats.[7,8,10]

Conversely, some life history traits of red-shoulders might suggest that they would not thrive in urban areas. Published accounts, particularly from the eastern half of North America, document a strong affinity of red-shouldered hawks for bottomlands, wetlands, and riparian areas,[10,11,12,13,14,15,16,17] a habitat association that might be limiting in urban areas. Indeed, some earlier reports indicate that red-shouldered hawks often nest in remote locations, away from human residences and disturbances such as roads.[12,18,19]

However, evidence from the field clearly shows that red-shouldered hawks have found ways to inhabit suburban areas in many parts of their range (figures 8.1, 8.2). Suburban red-shoulders have been deliberately studied in three locations: southern California,[9,20,21] central California,[22] and southwestern Ohio (figure 8.3, color plate 3, color plate 4).[14,15,23,24,25,26] Red-shouldered hawk nests

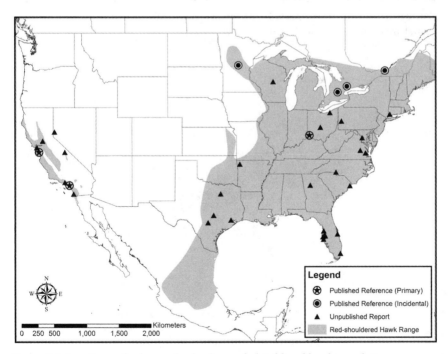

Figure 8.1. Locations of urban and suburban red-shouldered hawk populations.

Figure 8.2. Red-shouldered hawks in suburban areas around Cincinnati, Ohio, use residential areas for nesting and foraging. Photo by Sara Miller.

Figure 8.3. Colored plastic leg bands on suburban hawks provide opportunities for learning about survival and movements. Local residents often assist researchers by sending photographs of banded birds in their neighborhoods. Photo by Ruhikant Meetei.

near residences and other human activity also are mentioned in the context of other studies in primarily rural areas.[27,28,29,30] However, raptor researchers and other sources indicate that red-shouldered hawks are present in many urban areas, and the geographic spread of these anecdotal reports suggests that red-shoulders may use these areas in most parts of their range, given suitable habitat, particularly in the southeastern United States (figure 8.1).[31]

Long-term studies of urban red-shouldered hawks, conducted with different subspecies in different parts of the United States, offer the opportunity to investigate this species' urban adaptability. Here, we review the life history of red-shouldered hawks, focusing on two overarching questions:

1. What traits allow red-shouldered hawks to inhabit urban areas?
2. How do urban red-shouldered hawks differ from those in more natural sites?

Behavioral Ecology

Broadly defined, "behavior" covers most aspects of a bird's life, from selecting habitat to foraging, brooding and feeding its young, and defending its territory and nest. Our discussion of red-shouldered hawk behavior therefore addresses habitat selection, foraging ecology, and nest defense.

SELECTING HABITAT

Red-shouldered hawks' flexibility and ability to adapt to different landscapes, essentially variations on the structure of woodlands, have allowed them to inhabit urban environs. In eastern North America, red-shoulders use both bottomland hardwood riparian areas and wooded swamps and also upland mixed deciduous-coniferous or deciduous forests, and their habitats are often described as extensive stands of mature forest.[28,30,32,33,34] In south-central Florida, red-shouldered hawks inhabit areas that are primarily grasslands and wetlands with limited forest (approximately 6 percent).[35] In the suburbs of Cincinnati in southwestern Ohio, red-shouldered hawks nest in residential areas that are interspersed with small pockets of native oak-hickory (*Quercus* spp., *Carya* spp.) and beech-maple (*Fagus grandifolia, Acer saccharum*) forests, with riparian areas characterized by sycamore (*Platanus occidentalis*). Compared to nest sites in a nearby rural forested area (Hocking Hills region in southeastern Ohio), nest sites in suburban Cincinnati differed very little, except that the nest trees were closer to houses (75 m on average) and were more likely to be surrounded by lawn (one-third of all nests).[14]

Hawks at both the suburban and rural sites chose mostly native nest trees that were taller and had larger diameters than what was randomly available in the area and were closer to water (33 m and 27 m, respectively) than randomly located trees.[14] Abundance of red-shouldered hawks in suburban and rural areas was positively correlated to the number of small ponds within the survey area,[15] suggesting an association with water also shown in studies of eastern red-shoulders in more natural areas.[18,32,36,37] Annual home ranges of suburban hawks consisted of 41 percent native forest and 50 percent suburban development, and breeding home ranges (90 ha) contained an average of 169 human residences.[23]

The western red-shouldered hawk (*B. lineatus elegans*), a bird of more arid habitats, lives in areas that are much less forested and have much less standing water than those used in eastern North America. This subspecies occupies riparian zones and oak woodlands but also residential areas in southern[20] and central California.[22] In southern California, areas used by urban red-shoulders included nonnative vegetation (particularly *Eucalyptus* spp., pine [*Pinus*] spp., and fan palms [*Washingtonia filifera*]), lawns, athletic fields, parking lots, buildings, and roads, and also a portion of natural habitat, where coast live oak (*Quercus agrifolia*), western sycamore (*Platanus racemosa*), and Goodding's black willow (*Salix gooddingii*) dominated.[20] Annual home ranges of these urban birds contained 54 percent urban woodland (mostly exotics) and another 20–25 percent buildings, roads, and water.[9,20] In the residential and industrial areas of the Santa Clara Valley in central California, red-shoulders nested in riparian zones dominated by Fremont cottonwood (*Populus fremontii*), western sycamore, and willows (*Salix* spp.) and also in the associated uplands, which contained sparse nonnative trees, including *Eucalyptus*, palms, and conifers.[22]

Nest tree species selected by urban red-shoulders vary by location (figure 8.4). In southern California, 38 percent of nest trees were nonnative species, primarily *Eucalyptus*, a tall species that provides a stable location for a nest.[20] Similarly, red-shoulders in central California built in nonnatives trees, primarily *Eucalyptus*.[22] Windbreaks of nonnative *Eucalyptus* that were planted in areas that previously lacked large trees expanded the available suitable breeding habitat for red-shoulders in California.[20,22] In contrast, suburban red-shouldered hawks in southwestern Ohio nested mainly in native species, including planted trees in yards. Rural-nesting red-shoulders in a comparison site nearby also used mainly native species, although a few nested in pines in a plantation in a national forest.[14] Two suburban pairs nested on the rooftops of residential

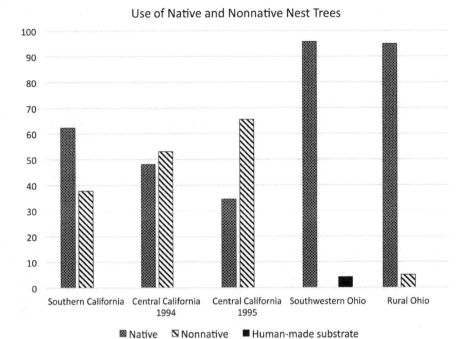

Figure 8.4. Use of native and nonnative nest trees (percent of total) by red-shouldered hawks in Ohio and California.

buildings (color plate 5) and one pair nested on a gas grill located on the deck of a home,[23,38] underscoring the adaptability of the species. Of hundreds of western territories, only one pair of the western subspecies also nested on an artificial structure, the lower crossmember of a high-tension power tower in Long Beach, California.

Home ranges of urban red-shoulders tend to be smaller than those of birds nesting in more natural areas, suggesting that urban areas provided abundant prey. Four urban birds in southern California had annual home ranges averaging 50 ha, compared to 113 ha for 12 birds in more natural habitat in the same region.[9] In suburban southwestern Ohio, annual home ranges averaged 165 ± 24 ha, and breeding birds occupied 90 ± 11 ha (n = 9 and 11, respectively),[23] which is considerably smaller than breeding home ranges measured for red-shoulders in more natural habitats of eastern North America.[7,8,10,23] Radio-tagged suburban birds used human-made structures within their home ranges, perching on utility poles, rooftops, and lawn furniture, and hunting at ornamental ponds and bird feeders.[20,23] Combined with the hawks' flexibility in using anthropogenic landscapes and nonnative trees, small home ranges probably contribute substantively to the species' ability to inhabit and thrive in urban areas.

Foraging Behavior

Red-shouldered hawks have catholic tastes and eat a variety of taxa, including small mammals (mice, voles, shrews), amphibians (primarily frogs in most areas), and reptiles (especially snakes), as well as occasional birds, fish, and invertebrates (mostly crayfish and earthworms).[6] This generalist diet is likely an important determinant of their ability to use our urban landscapes. Populations in the northern part of the range eat more small mammals, whereas those in the southern part tend to feed on amphibians and reptiles,[6,39] and local diets may shift from year to year, depending on availability.[40] Suburban red-shouldered hawks apparently capture typical prey for their latitude, although no direct comparisons of suburban and nearby rural populations have been undertaken. Diet of Cincinnati birds determined by direct observations included 32 percent small mammals, 23 percent reptiles, 19 percent invertebrates, 18 percent amphibians, 6.9 percent birds, and 2.5 percent fish.[25]

Anecdotal evidence indicates that urban red-shoulders scavenge human-provided food on occasion. Suburban birds have been observed eating waste food around buildings and athletic fields[20] and on an open compost pile, as well as cat food (tuna) placed by a building. They've also feasted on pizza, beans or chili, and koi from ornamental ponds. One indulgent landowner regularly provided raw chicken for "her" nesting pair, one of which was among the longest-lived, most productive females in the Cincinnati study. Another creative landowner regularly attached meat to a fishing line (without a hook), cast it into his yard, then reeled it in, with the resident red-shoulders following the treat almost to his feet before claiming it.

Nest Defense and Other Behaviors

One important criterion for the successful occupation of urban environments is the ability to adapt to the close proximity of humans,[1] which implies a certain amount of behavioral flexibility, as described in chapter 2.[3,4,5] Urban red-shouldered hawks behave rather differently than their rural counterparts. Individuals tolerate human disturbances around their nests, ranging from athletic competitions to walkers/runners and to homeowners mowing lawns. Researchers on foot often can approach suburban birds to within 25 m, close enough to read color-bands using binoculars.[20] Incubating or brooding birds only rarely flushed from a suburban nest in response to a researcher walking up to the nest tree, whereas in the rural comparison area, approximately one-third of birds flushed silently and flew away from the observer.

As illustrated in a case study in chapter 17, urban red-shoulders occasionally become defensive of their nests, diving at or hitting residents on the ground or researchers climbing to nests. Such attacks can cause injuries but most often are simply startling, as when defensive hawks snatched and carried off a baseball cap and a child's sequined headband directly from their owners' heads. Defensive diving behavior is rare at rural nests; researchers climbing to suburban nests in Cincinnati were hit by hawks on 11 percent of 166 climbs but only approximately 2 percent of more than 200 climbs to rural nests. Most suburban residents near aggressive hawks tolerated them and coped by avoiding the nest area (figure 8.5), wearing a helmet or hard hat, or carrying an umbrella. In a few cases, where hawks repeatedly injured residents, particularly children or the elderly, we removed nests before egg laying, removed the young and fostered them in other nests (with appropriate permits), or captured and relocated the adults.[20]

Figure 8.5. Warning sign posted by apartment management near an aggressive red-shouldered hawk's nest in Cincinnati, Ohio. Photo by Melinda Simon.

Population Ecology

REPRODUCTION

Urban red-shouldered hawks reproduce at rates similar to those of individuals nesting in more natural areas. In a large 19-year study, Cincinnati red-shoulders produced 1.55 ± 0.04 young per active nest (i.e., nest with eggs), compared to 1.54 ± 0.08 young per active nest in the nearby rural Hocking Hills study area; similarly, suburban birds also did not differ from rural ones in terms of the number of young produced per successful nest (i.e., nest where at least one young fledged; 2.59 ± 0.04 vs. 2.61 ± 0.06, respectively; average of annual means, 1997–2015).[6,14,24] Reproductive rates varied significantly among the suburban nesting territories, with the most productive 25 percent of the territories producing 44 percent of the nestlings and the least productive 25 percent producing only 7 percent of the young, suggesting some inherent variability in the habitat quality, likely attributable to prey abundance, predator density, or both.[24] When compared to the reproductive rates measured in natural areas across the red-shouldered hawk's range,[6] the reproductive rates of Ohio birds were representative. Comparisons among studies should be made with caution, however, as researchers measure and report productivity in slightly different ways, and environmental, habitat, and interannual variability also likely affect productivity.[35]

In southern California, urban red-shouldered hawks produced 1.80 young per nesting attempt or 2.50 young per successful nest.[20] This rate was greater than that measured by another researcher in the same area,[41] although methods were not entirely comparable.

In central California, reproduction of urban red-shouldered hawks averaged 1.6–1.8 young fledged per nest (not defined) and 2.0–2.3 young fledged per successful nest. Hawks nesting in exotic tree species raised significantly more young per nest than those using native species in the first year of the study but not in the second.[22]

SURVIVAL

Urban red-shouldered hawks fall victim to a variety of predators and to human-made hazards such as collisions with cars and electric lines, as discussed in chapters 4 and 14. Great horned owls (*Bubo virginianus*) are the most important predators in suburban southwestern Ohio, where they killed 14 of 28 nestlings that died. Raccoons (*Procyon lotor*) killed some nestlings, and others simply fell from their nests.[42] Somewhat surprisingly, eastern gray squirrels (*Sciurus carolinensis*) contesting red-shoulders for possession of the nest were responsible for some nest failures. One squirrel attacked a brooding female, causing her to abruptly spring up and one nestling to be ejected from the nest.[42] Great horned owls also can kill incubating females.[42] In natural areas, great horned owls and raccoons also are blamed for predation of nestlings, but in most cases, predation events were not observed or filmed, and evidence was circumstantial.[43,44]

For some small raptors, the risk of predation is reduced in urban habitats compared to more natural habitats.[2] However, because the great horned owl is a generalist predator that is apparently thriving in some urban areas[2,45] and raccoon populations may actually increase in urban settings,[46] release from predation pressure seems unlikely, at least in suburban southwestern Ohio. Conversely, despite entry into several hundred great horned owl nests in coastal southern California, no evidence of predation by this species on red-shouldered hawks has been observed there.

It is not clear whether urban red-shouldered hawks suffer higher mortality than those inhabiting more natural areas. Limited evidence from banding data (recoveries) suggests survival is similar. In both suburban and rural areas of Ohio, the mean age at death was about 2 years old; 50 percent of hawks banded as nestlings were dead by age 1.1–1.2 years, and 95 percent by age 5 years old.[26]

The North American longevity record for a red-shouldered hawk is that of a female banded as a nestling, recaptured when 10 years of age, and resighted often. During her 26 years, she had at least two long-term mates and at least one extra-pair copulation.[47] She nested in coast live oak, black willow, western sycamore, and a California fan palm, all while thousands of United State Marines trained throughout her territory and under her nest trees.

Humans and human-made structures may be significant causes of mortality for urban raptors,[48] as shown in chapter 14. Analysis of banding recoveries for suburban red-shoulders indicated that, of the birds for whom cause of death was known, 38 percent were killed by collision with motor vehicles and 31 percent by electrocution on power lines or electric fences, although sample sizes were small (*n* = 13 with known cause of death).[26] Two hawks, a male and a female, were found electrocuted together beneath a utility pole, where they were probably copulating or at least touching each other; their electrocutions caused a local power outage. In southern California, the majority (64 percent) were killed by collision with motor vehicles, with one being poisoned and another oiled (*n* = 11 with known cause of death).

Dispersal

Raptors may become urban by at least two different paths: they may persist in a rural area that is overtaken by suburban sprawl, or they may colonize an urban area that has become suitable, particularly if new habitat is created there.[2,20] Further, the ability to colonize new areas may be influenced by the extent to which a species is able to disperse.[4]

Generally, red-shouldered hawks do not disperse very far. Mean distance of banded red-shouldered hawks from their natal nest when found was 39 km in southwestern Ohio[26] and 55 km in southern California. A few individuals in each area traveled more than 100 km. In southwestern Ohio, five red-shoulders were found 103–500 km from their nests, whereas in southern California, 10 were found 111–843 km from their nests (color plate 6).[21] The three southern California birds that traveled farthest (374, 804, and 843 km) were considered vagrants because they were found outside the known range of the species.[21] The ability to disperse relatively long distances in some cases and the potential for vagrancy may allow red-shouldered hawks to colonize new areas, including urban areas, and be successful there. This may be an especially important trait as natural habitat continues to be lost to urbanization or shifts due to a changing climate.

Conservation and Management

Behavioral flexibility and several life history traits have allowed red-shouldered hawks to inhabit urban areas in many parts of their distribution. Rangewide, their populations have tended to increase or remain stable over recent decades.[49,50] Yet due to the limited number of studies in urban areas and the diversity of habitats and cities in which red-shoulders live, their overall conservation status in cities remains unclear. Because of the broad latitudinal and longitudinal distribution of red-shouldered hawks, the cities within their range vary widely in native habitat type, prey abundance, predator density, and anthropogenic threats; thus, we expect that red-shoulder population trends in those cities may vary also.

In suburban southwestern Ohio, red-shouldered hawks are thriving. They are as successful as the nearby rural hawks, choosing equivalent nest trees, reproducing at the same rate, and finding typical prey species. Yet some territories have been lost to continued suburban sprawl: of a sample of 22 territories where red-shouldered hawk nestlings were banded in 1963–77, only 10 still contained hawks by 1997–98.[14] It is possible that these losses may be offset by ongoing adaptation by the hawks, possibly resulting in previously unsuitable sites becoming usable.

In southern California, the number of occupied red-shouldered hawk breeding territories, both rural and urban, has plummeted over the last decade. The severity of an extended drought may be playing a role, but evidence also suggests secondary poisoning by anticoagulants (rodent poison) and West Nile virus may have contributed to the decline.

Recommendations for conservation of red-shouldered hawks in urban areas differ little from general strategies for conserving wildlife in cities. We recommend preserving natural areas that produce prey and allow for nesting, including forests, old fields, and wild edges around streams, ponds, and other wetlands. Likewise, we recommend eliminating use of chemicals that reduce prey abundance or directly or indirectly threaten top-level predators. Importantly, as demonstrated in chapters 15 and 17, continuing education of residents will both help reduce human-hawk conflicts and facilitate an appreciation for wildlife in general.

Additional research is needed to better understand the ecology of urban red-shouldered hawks, particularly in little-studied portions of the species' range, such as southern Florida, the northern edge of the range, and newly colonized areas. Fortunately, compared to many raptors, urban red-shouldered hawks are conveniently easy to study due to their noisy courtship, visible nests, and relatively high densities. As a generalist species living in suburban regions across many areas of North America, the red-shouldered hawk is an excellent adaptable top predator model on which to study the effects of human-dominated environments. Finally, as an added bonus, this highly charismatic species provides numerous opportunities for researchers to interact with local residents and provide accurate information about the ecology of native wildlife, a key component in urban conservation efforts.

Acknowledgments

Southern California field data collection was enhanced by the presence of Charles Beranek, Jeremy Blatchford, Pete DeSimone, Richard Jackson, Susan Gallaugher, Judy Henckel, Ed Henckel, Jeff Kidd, Donna Krucki, Jim Luttrell, Linda Luttrell, Mike Marks, Karly Moore, Donna O'Neill, Richard O'Neill, Chris Niemela, Joe Papp, Scott Thomas, Zack Smith, and Michael van Hattem. We also thank Jeff Hays, Melinda Simon, Ann Wegman, Sara Miller, and Sandra Stone in southern Ohio for their assistance with fieldwork and data collection.

Literature Cited

1. Luniak, M. 2004. "Synurbization—Adaptation of Animal Wildlife to Urban Development." In *Proceedings of the 4th International Urban Wildlife Symposium*, edited by W. W. Shaw, K. L. Harris, and L. Van Druff, 50–55. Tucson: University of Arizona.

2. Chace, J. F., and J. J. Walsh. 2006. "Urban Effects on Native Avifauna: A Review." *Landscape and Urban Planning* 74:46–69.

3. Ditchkoff, S. S., S. T. Saalfeld, and C. J. Gibson. 2006. "Animal Behavior in Urban Ecosystems: Modifications Due to Human-Induced Stress." *Urban Ecosystems* 9:5–12.

4. Møller, A. P. 2009. "Successful City Dwellers: A Comparative Study of the Ecological Characteristics of Urban Birds in the Western Palearctic." *Oecologia* 159:849–58.

5. Carrete, M., and J. L. Tella. 2011. "Inter-Individual Variability in Fear of Humans and Relative Brain Size of the Species Are Related to Contemporary Urban Invasion in Birds." *PLoS ONE* 6: e18859.

6. Dykstra, C. R., J. L. Hays, and S. C. Crocoll. 2008. "Red-Shouldered Hawk (*Buteo lineatus*)." In *The Birds of North America*, edited by P. G. Rodewald. Ithaca: Cornell Lab of Ornithology. Accessed April 18, 2017. https://birdsna.org/Species-Account/bna/species/reshaw.

7. Parker, M. A. 1986. "The Foraging Behavior and Habitat Use of Breeding Red-Shouldered Hawks in Southeastern Missouri." MS thesis, University of Missouri.

8. Senchak, S. S. 1991. "Home Ranges and Habitat Selection of Red-Shouldered Hawks in Central Maryland: Evaluating Telemetry Triangulation Errors." MS thesis, Virginia Polytechnic Institute and State University.

9. Bloom, P. H., M. D. McCrary, and M. J. Gibson. 1993. "Red-Shouldered Hawk Home-Range and Habitat Use in Southern California." *Journal of Wildlife Management* 57:258–65.

10. Howell, D. L., and B. R. Chapman. 1997. "Home Range and Habitat Use of Red-Shouldered Hawks in Georgia." *Wilson Bulletin* 109:131–44.

11. Portnoy, J. W., and W. E. Dodge. 1979. "Red-Shouldered Hawk Nesting Ecology and Behavior." *Wilson Bulletin* 91:104–17.

12. Bosakowski, T., D. G. Smith, and R. Speiser. 1992. "Status, Nesting Density, and Macrohabitat Selection of Red-Shouldered Hawks in Northern New Jersey." *Wilson Bulletin* 104:434–46.

13. Moorman, C. E., and B. R. Chapman. 1996. "Nest-Site Selection of Red-Shouldered and Red-Tailed Hawks in a Managed Forest." *Wilson Bulletin* 108:357–68.

14. Dykstra, C. R., J. L. Hays, F. B. Daniel, and M. M. Simon. 2000. "Nest Site Selection and Productivity of Suburban Red-Shouldered Hawks in Southern Ohio." *Condor* 102:401–8.

15. Dykstra, C. R., F. B. Daniel, J. L. Hays, and M. M. Simon. 2001. "Correlation of Red-Shouldered Hawk Abundance and Macrohabitat Characteristics in Riparian Zones." *Condor* 103:652–56.

16. McLeod, M. A., B. A. Belleman, D. E. Andersen, and G. Oehlert. 2000. "Red-Shouldered Hawk Nest Site Selection in North-Central Minnesota." *Wilson Bulletin* 112:203–13.

17. Balcerzak, M. J., and P. B. Wood. 2003. "Red-Shouldered Hawk (*Buteo lineatus*) Abundance and Habitat in a Reclaimed Mine Landscape." *Journal of Raptor Research* 37:188–97.

18. Bednarz, J. C., and J. J. Dinsmore. 1981. "Status, Habitat Use, and Management of Red-Shouldered Hawks in Iowa." *Journal of Wildlife Management* 45:236–41.

19. Bosakowski, T., and D. G. Smith. 1997. "Distribution and Species Richness of a Forest Raptor Community in Relation to Urbanization." *Journal of Raptor Research* 31:26–33.

20. Bloom, P. H., and M. D. McCrary. 1996. "The Urban Buteo: Red-Shouldered Hawks in Southern California." In *Raptors in Human Landscapes: Adaptations to Built and Cultivated Environments*, edited by D. Bird, D. Varland, and J. J. Negro, 31–39. San Diego: Academic Press.

21. Bloom, P. H., J. M. Scott, J. M. Papp, S. E. Thomas, and J. W. Kind. 2011. "Vagrant Western Red-Shouldered Hawks: Origins, Natal Dispersal Patterns, and Survival." *Condor* 113:538–46.

22. Rottenborn, S. C. 2000. "Nest-Site Selection and Reproductive Success of Urban Red-Shouldered Hawks in Central California." *Journal of Raptor Research* 34:18–25.

23. Dykstra, C. R., J. L. Hays, F. B. Daniel, and M. M. Simon. 2001. "Home Range and Habitat Use of Suburban Red-Shouldered Hawks in Southwestern Ohio." *Wilson Bulletin* 113:308–16.

24. Dykstra, C. R., J. L. Hays, and M. M. Simon. 2009. "Spatial and Temporal Variation in Red-Shouldered Hawk Reproductive Rates." *Condor* 111:177–82.

25. Dykstra, C. R., J. L. Hays, M. M. Simon, and F. B. Daniel. 2003. "Behavior and Prey of Nesting Red-Shouldered Hawks in Southwestern Ohio." *Journal of Raptor Research* 37:177–87.

26. Dykstra, C. R., J. L. Hays, M. M. Simon, J. B. Holt, Jr., G. R. Austing, and F. B. Daniel. 2004. "Dispersal and Mortality of Red-Shouldered Hawks Banded in Ohio." *Journal of Raptor Research* 38:304–11.

27. Campbell, C. A. 1975. "Ecology and Reproduction of Red-Shouldered Hawks in the Waterloo Region, Southern Ontario." *Journal of Raptor Research* 9:12–17.

28. Morris, M. M. J., and R. E. Lemon. 1983. "Characteristics of Vegetation and Topography near Red-Shouldered Hawk Nests in Southwestern Quebec." *Journal of Wildlife Management* 47:138–45.

29. Dent, P. 1994. "Observations on the Nesting Habits of Red-Shouldered Hawks in York Region." *Ontario Birds* 12:85–94.

30. Henneman, C. 2006. "Habitat Associations of Red-Shouldered Hawks in Central Minnesota Landscapes." MS thesis, University of Minnesota.

31. Wheeler, B. K. 2003. *Raptors of Eastern North America*. Princeton: Princeton University Press.

32. Titus, K., and J. A. Mosher. 1981. "Nest-Site Habitat Selected by Woodland Hawks in the Central Appalachians." *Auk* 98:270–81.

33. Bednarz, J. C., and J. J. Dinsmore. 1982. "Nest-Sites and Habitat of Red-Shouldered and Red-Tailed Hawks in Iowa." *Wilson Bulletin* 94:31–45.

34. Dijak, W. D., B. Tannenbaum, and M. A. Parker. 1990. "Nest-Site Characteristics Affecting Success and Reuse of Red-Shouldered Hawk Nests." *Wilson Bulletin* 102:480–86.

35. Morrison, J. L., M. McMillian, J. B. Cohen, and D. H. Catlin. 2007. "Environmental Correlates of Nesting Success in Red-Shouldered Hawks." *Condor* 109:648–57.

36. Armstrong, E. and D. Euler. 1983. "Habitat Usage of Two Woodland Buteo Species in Central Ontario." *Canadian Field-Naturalist* 97:200–207.

37. Woodrey, M. S. 1986. "Characteristics of Red-Shouldered Hawk Nests in Southeast Ohio." *Wilson Bulletin* 98:466–69.

38. Hays, J. L. 2000. "Red-Shouldered Hawks Nesting on Human-Made Structures in Southwest Ohio." In *Raptors at Risk: Proceedings of the V World Conference on Birds of Prey and Owls*, edited by R. D. Chancellor and B.-U. Meyburg, 469–71. Berlin, Germany: World Working Group on Birds of Prey and Owls; Surrey, BC: Hancock House Publishers.

39. Strobel, B. N., and C. W. Boal. 2010. "Regional Variation in Diets of Breeding Red-Shouldered Hawks." *Wilson Journal of Ornithology* 122:68–74.

40. Bednarz, J. C., and J. J. Dinsmore. 1985. "Flexible Dietary Response and Feeding Ecology of the Red-Shouldered Hawk, *Buteo lineatus*, in Iowa." *Canadian Field-Naturalist* 99:262–64.

41. Wiley, J. W. 1975. "The Nesting and Reproductive Success of Red-Tailed Hawks and Red-Shouldered Hawks in Orange County, California, 1973." *Condor* 77:133–39.

42. Miller, S. J., C. R. Dykstra, M. M. Simon, J. L. Hays, and J. C. Bednarz. 2015. "Causes of Mortality and Failure at Suburban Red-Shouldered Hawk (*Buteo lineatus*) Nests." *Journal of Raptor Research* 49:152–60.

43. Crocoll, S. T., and J. W. Parker. 1989. "The Breeding Biology of Broad-Winged and Red-Shouldered Hawks in Western New York." *Journal of Raptor Research* 23:125–39.

44. Townsend, K. A. L. 2006. "Nesting Ecology and Sibling Behavior of Red-Shouldered Hawks at the St. Francis Sunken Lands Wildlife Management Area in Northeastern Arkansas." MS thesis, Arkansas State University.

45. Holt, J. B., Jr. 1996. "A Banding Study of Cincinnati Area Great Horned Owls." *Journal of Raptor Research* 30:194–97.

46. Prange, S., S. D. Gehrt, and E. P. Wiggers. 2003. "Demographic Factors Contributing to High Raccoon Densities in Urban Landscapes." *Journal of Wildlife Management* 67:324–33.

47. McCrary, M. D., and P. H. Bloom. 1984. "Observations on Female Promiscuity in the Red-Shouldered Hawk." *Condor* 86:486.

48. Hager, S. B. 2009. "Human-Related Threats to Urban Raptors." *Journal of Raptor Research* 43:210–26.

49. Sauer, J. R., J. E. Hines, and J. Fallon. 2005. "The North American Breeding Bird Survey, Results and Analysis 1996–2005." Version 6.2.2006. Patuxent Wildlife Research Center, Laurel, MD.

50. Bildstein, K. L., J. P. Smith, and E. Ruelas Inzuena. 2008. "The State of North American Birds of Prey." Series in Ornithology No. 3. American Ornithologists' Union and Nuttall Ornithological Club.

CHAPTER 9

Harris's Hawks: All in the Family

Clint W. Boal and James F. Dwyer

F ROM PHOENIX, ARIZONA, SOUTH TO Buenos Aires, Argentina, residents of many cities have Harris's hawks (*Parabuteo unicinctus*) as neighbors. This intriguing raptor has adapted to human presence in many urban landscapes, generally without developing conflicts with human residents. Harris's hawks are often popular among nature enthusiasts due to the species' accessibility, large size, and easily recognizable plumage. Harris's hawks have also garnered popularity because of their unique ecology as the proverbial wolves of the air; like a wolf pack, a family group of Harris's hawks will work together to raise young and will hunt cooperatively.[1,2]

An adult Harris's hawk is a visually striking bird (color plate 7), with an overall dark chocolate-brown plumage, distinct rufous feathers on the upper-and underwing coverts and the thighs, bright white undertail coverts, and a black tail with white at the base and trailing edge. When Harris's hawks are perched, the rufous wing coverts give the appearance of dark-red shoulders for which they are sometimes called "bay-winged hawks."[3] Their scientific genus *Parabuteo* means "near buteo," and as the name suggests, Harris's hawks are a somewhat unusual species. In most ways, they appear like typical buteonine hawks but with wings that are slightly shorter than those of similar-sized buteo hawks and a long tail more characteristic of an accipiter. The combination of these features makes

Harris's hawks quick and agile in flight and capable of capturing a wide variety of prey ranging in size from songbirds to jackrabbits.[4]

It is not clear exactly when Harris's hawks first began occupying urban landscapes; indeed, they were traditionally considered intolerant of human presence, especially near nest sites.[5,6] Thus it was somewhat surprising that some Harris's hawks were found occupying Tucson, Arizona, in the mid-1970s[5] and that the population had apparently increased by the mid-1980s.[7] Similarly, the species was seen around Lima, Peru, in the mid-1980s and had established a resident population by 2000,[8] though the population was composed, at least in part, of escaped or released falconry birds. Today, it is not surprising to find Harris's hawks as residents in many cities across the species' distribution.

Behavioral Ecology

SOCIAL BEHAVIOR

Unlike most birds of prey, Harris's hawks are often highly social. This sociality includes both their reproductive ecology and their hunting behavior. Many Harris's hawks live in "family" groups consisting of the breeding "alpha" pair and multiple adult or juvenile "beta" and "gamma" helpers that may or may not be related. Harris's hawks demonstrate their rank within the family group by perching together on a single structure, with the most dominant alpha birds perching highest and the most subordinate gamma birds perching lowest.[4]

In addition to typical monogamy in which the dominant or alpha male and female mate, a family group may engage in polyandry, in which more than one male mates with the alpha female.[9] More rarely, polygyny, in which two alpha females mate with the same male and both females contribute eggs to the nest, may occur.[10] The beta and gamma members of the family group assist in foraging, provisioning of nestlings, and territory defense, and some beta males provide direct care to eggs and young.[11]

In exurban areas, Harris's hawks will sometimes aggregate at water bodies, particularly during nonbreeding seasons; indeed, this association with water has been interpreted as being important for allowing the species to persist in desert landscapes.[1] However, this does not preclude family groups from defending their territories year-round. Territorial conflicts can involve a series of escalating displays that rarely, but may eventually, result in physical altercations.[12] These appear to be consistent between urban and exurban areas.

Habitat

Across their substantial distribution, Harris's hawks generally occupy open country, deserts, brushlands, and scrub woodland ranging from sea level to as high as 1,900 m; at higher elevations in Chile, the species can be found in mixed deciduous woodlands.[4,13,14] In Arizona and New Mexico, Harris's hawks are typically found in the arid deserts and savannah grasslands of the Sonoran and Chihuahuan ecoregions. Southeast into Texas, Harris's hawks occupy portions of the Chihuahuan desert, the oak savannah of the Edwards Plateau, brushlands of the South Texas Plains, and the Gulf Coast Prairies.

Harris's hawks' use of urban areas, where they often nest in evergreen trees, is a departure in many ways from their use of exurban areas. In the arid desert and brushland landscapes of southern Arizona and South Texas, cities can be oases of taller, nonnative tree species, with ample hunting perches provided by utility poles and other structures, and abundant presence of water due to landscape and yard irrigation, water features, and pools. Harris's hawks use these resources and often hunt in the remnants of natural vegetation remaining in open spaces such as rights-of-way and drainages.

Nest-Site Selection

Nest-site selection differs between exurban and urban environments. In exurban environments, depending on ecoregion, Harris's hawks typically nest in paloverde (*Cercidium* spp.), saguaros (*Carnegiea gigantean*), oaks (*Quercus* spp.), hackberry (*Celtis* spp.), and mesquites (*Prosopis* spp.). Of 306 nests in Arizona, 75 percent in exurban areas were in saguaro cactus and 20 percent were in paloverde trees.[15] Most often, these nests ranged from about 4 to 6 m above the ground. In contrast, the average nest height was 12.5 m for 38 nests in urban Tucson, Arizona, where the majority of nests were placed in substantially taller, introduced trees.[7] Of 72 urban Harris's hawk nests monitored, 81 percent were in introduced species, 68 percent of which were pines (*Pinus* spp.; figure 9.1) and *Eucalyptus* species.[7] Urban Harris's hawks also appear to have a relatively high tolerance of human activities, with nests averaging 21.2 m from the nearest occupied domicile and 20.8 m from areas of consistent human activity (e.g., parking areas, patios, etc.).[7]

PLATE 1. Mississippi kites have become common urban residents throughout much of the Southern Great Plains due to their range expansion associated with European settlement and agroforestry practices. *Photograph courtesy of Clint Boal.*

PLATE 2. Cooper's hawks may be year-round or seasonal residents in urban areas, depending on the location. This Cooper's hawk nested successfully in an urban area of Milwaukee, Wisconsin, and wintered in a suburban area of McGregor, Texas, for approximately three months. *Photo by Natalie Kutach (McGregor, Texas), courtesy of William Stout.*

PLATE 3. Red-shouldered hawks in suburban areas around Cincinnati, Ohio, use residential areas for nesting and foraging. Colored plastic leg bands on suburban hawks provide opportunities for learning about survival and movements. Local residents often assist researchers by sending photographs of banded birds in their neighborhoods. *Photograph by Murielle Bennett, courtesy of Melinda Simon.*

PLATE 4. This red-shouldered hawk was banded as a nestling in a suburb of Cincinnati, Ohio, in 2011. Subsequently, it was photographed multiple times from 2013–2016 by a local photographer/birder at a site 17 km from its natal nest. *Photograph courtesy of Ruhikant Meetei.*

PLATE 5. Red-shouldered hawks nested on the rooftop of a three-story suburban apartment building in Cincinnati, Ohio. Although most urban raptors nest in trees, they will occasionally use man-made structures for nests. *Photograph courtesy of Cheryl Dykstra.*

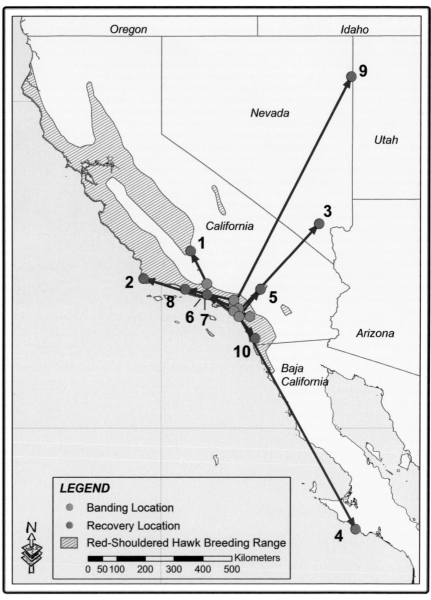

PLATE 6. Most red-shouldered hawks banded as nestlings disperse short distances from their natal nests (median distance = 24 km in southern California). However, a small percentage disperse much further. Ten red-shouldered hawks (8.4% of the total) banded in urban and rural nests in southern California dispersed more than 100 km, with three of them found outside of the known breeding range. Each of these long-distance dispersers is represented by a blue track between the banding location (green circle) and the recovery location (red circle). *(Published in Condor 113:538–546 and reprinted with permission.)*

PLATE 7. Harris's hawks are quick and agile in flight, and capable of capturing a wide variety of prey species, which may facilitate their occupancy of urban landscapes. *Photograph courtesy of Steve Wolfe.*

PLATE 8: Barred owls are a common nocturnal denizen of urban landscapes across eastern North America. This adult female barred owl is guarding her nest cavity in a suburban yard in Charlotte, North Carolina. *Photograph courtesy of Richard O. Bierregaard.*

PLATE 9. Powerful owls are the largest owl in Australia. Some powerful owls are attracted to urban areas that have ample prey, such as the common brushtail possum this male powerful owl captured in an urban reserve. A conservation issue, however, is that urban areas often lack trees suitably large enough to have cavities for the owls to nest in. *Photograph by John White.*

PLATE 10. Burrowing owls are unique raptors in that they nest in burrows dug by small mammals such as prairie dogs and rabbits. *Photograph by Rob Palmer.*

PLATE 11. Peregrine falcons have become common inhabitants of cities, nesting on tall buildings and other structures. This individual nests at City Hall in San Jose, California. *Photograph by Nick Dunlop.*

Plate 12. Urban raptors are exposed to a variety of hazards they don't normally encounter in natural areas. This great horned owl became entrapped in a soccer net and required admission to an urban rehabilitation facility.

PLATE 13. Methods used for raptor conservation programming frequently include presentation of live raptors. Many of these birds, often referred to as 'wildlife ambassadors' have been injured and rehabilitated, but are impaired and could not survive if released. *Photograph courtesy of Amber Burnette.*

PLATE 14. Many urban raptors, especially woodland raptors such as this juvenile Cooper's hawk, can become quite acclimated to human presence and activities, and readily take advantage of resources provided in urban environments. *Photograph courtesy of Doris Evans.*

PLATE 15. The majority of information on urban raptors is related to nesting and productivity due to the relative ease of collecting data at known nest sites compared to observing behaviors away from the nest. Urban raptors, such as this adult female and three fledgling western screech-owls from a backyard in Tucson, Arizona, can become quite approachable and will take advantage of features such as this water dish. *Photograph courtesy of Doris Evans.*

PLATE 16. Some raptor species may influence the presence and distribution of other raptor species. Due to their size and nocturnal habits, great horned owls will both usurp existing nests and depredate nestlings and adults of smaller raptors. During daylight, however, smaller, more agile raptors, such as the Cooper's hawk pictured here, sometimes attack owls to try to drive them from the area. *Photograph courtesy of Kim Domina.*

Figure 9.1. Harris's hawks often nest in introduced tree species planted in urban areas, such as this eldarica pine (*Pinus eldarica*) in Tucson, Arizona. Photograph courtesy of James F. Dwyer.

HOME RANGES

The home-range size for Harris's hawks in exurban areas in New Mexico was estimated to be 5.5 km².[16] No similar data are available for urban birds, but breeding territories appear to be relatively stable in both exurban and urban settings.[7,17]

DIET

Harris's hawks are generalist predators, capable of capturing a wide array of prey. Mammals—especially cottontail rabbits (*Sylvilagus* spp.), ground squirrels (*Sciuridae* spp.), and woodrats (*Neotoma* spp.)—account for more than 70 percent of their prey biomass in Arizona and New Mexico.[14] Other prey include a wide variety of birds and occasionally reptiles.[4,14] Presumably urban Harris's hawks hunt similarly, but little diet information has been compiled for the species in urban habitats. An increased reproductive rate in Tucson, Arizona, was believed to have been facilitated by increased prey availability compared to the exurban desert habitat.[7] In Lima, Peru, Harris's hawks capture large numbers of Pacific doves (*Zenaida meloda*),[8] possibly increasing their hunting success of this otherwise challenging prey species by ambushing them at urban bird feeders. However, urbanization does not necessarily equate to an increase in native prey; the negative influence of urbanization on native species such as the degu (*Octodon degus*) resulted in Harris's hawks shifting their diet to the introduced, and urban-favoring, European rabbit (*Oryctolagus cuniculus*) in Chile.[18]

COOPERATIVE HUNTING

One of the fascinating aspects of Harris's hawk foraging is that they can be cooperative hunters, working together to capture and share large prey. Cooperative hunts can take different forms, from multiple hawks converging on an exposed prey animal to long "relay" chases where the lead position is changed among hawks.[1,19] These latter chases may include interception flights where one hawk tries to cut off an anticipated escape route while others are chasing the prey.[1] Alternatively, when one hawk makes an attack, another may ascend upward to gain height and facilitate a follow-up attack if the first hawk misses.[1] If a family group has a prey animal surrounded in dense cover like a large thorny cactus, one member of a hunting group will crawl into the cover to flush prey into escape routes being watched by other group members. Cooperative hunting is common in the northern part of the species' range, where larger groups of hunting Harris's hawks have greater success in capturing prey.[1,19] Prey size, difficulty in capturing prey, or the limited hours of the day when prey are active during the hot summers of the arid southwest may favor cooperative hunting in the northern part of the species' range.[19,20,21] Unfortunately, there are no data comparing group size to hunting success in urban areas. Given the premise of increased prey abundance and a presumably less challenging landscape to hunt in, it may be that cooperative hunting is not as advantageous in urban areas. This is circumstantially supported by the observation that average group size was 3.8 hawks among exurban nests around Tucson, Arizona, compared to 2.1 hawks within nearby urban areas.[7] Comparison of group size to hunting success more broadly across the species' range would be informative.

Population Ecology

Reproduction

When comparing Harris's hawks in exurban and urban areas, researchers found urban nesting apparently provides some advantages in terms of reproductive success. Harris's hawks in exurban settings usually only produce one brood of nestlings per year, though second and even third broods occur occasionally.[4] In one study comparing 19 exurban and 46 urban Harris's hawk family groups in Arizona,[7] exurban groups only produced one brood per year and fledged 1.74 young per successful nest. In contrast, of 44 successful urban groups, 59 percent produced one brood, 29 percent produced two broods, and 11 percent produced three broods annually. Additionally, the urban groups fledged an average of 2.21 young per successful nest and, due to multiple broods, 3.45 young per family group per year.

Survival

Causes of mortality among Harris's hawks are not well understood. Harris's hawks respond aggressively to great horned owls (*Bubo virginianus*) and coyotes (*Canis latrans*), suggesting they are perceived as predation risks;[1] indeed, mortality by great horned owls has been documented.[4,11] Harris's hawks' affinity for water, at least in desert regions, has also led to some drowning in stock tanks.[15] Recently fledged juveniles may also become impaled on cacti.[4,15] Leg-hold traps, box traps, and poisoned baits can be mortality factors in some exurban areas.[15,21] Mortality by persecution occurs in both urban[7] and exurban areas.[4]

Urban landscapes present mortality factors for Harris's hawks that they likely do not encounter, or encounter at much reduced rates, in exurban settings. These mortality factors include disease (e.g., trichomoniasis),[22] drowning in swimming pools, secondary poisoning by consuming poisoned animals, collisions with windows, and collisions with cars.[23,24] However, these are relatively inconsequential in comparison to electrocutions (figure 9.2), which are a primary cause of injury and mortality and a major conservation concern for Harris's hawks in urban areas.[24,25,26] Electrocution disproportionately affects adult females and recently fledged juveniles. This is attributed to a combination of adult females transferring prey to juveniles and the clumsiness of juveniles learning to fly; these actions increase the potential of a hawk (or multiple hawks in contact with each other) contacting the differentially energized conductors on utility poles or transformers.[24] Power companies in some cities like Tucson

Figure 9.2. Electrocutions are the primary identified cause of mortality of Harris's hawks, such as the victim on top of this transformer, in urban areas. This cause of mortality disproportionately affects females and juvenile hawks; when the female is exchanging prey with juveniles on a power pole or transformer, there is an increased risk of hawks contacting energized conductors. Photograph courtesy of James F. Dwyer.

and Phoenix, Arizona, are engaging in active conservation efforts by installing protective covers on energized equipment to alleviate electrocution mortality of Harris's hawks (figure 9.3). These conservation efforts also reduce associated legal and financial costs of power outages sometimes caused by avian electrocutions, benefiting human residents as well as Harris's hawks. Undoubtedly, other raptor species also benefit from these conservation actions.

IMMIGRATION, EMIGRATION, AND DISPERSAL

Little is known about dispersal patterns of Harris's hawks.[7] However, five hawks marked at exurban nests dispersed an average of 72 km (20.9 standard error [SE]) into Tucson, where three became breeders and two became helpers in family groups. In contrast, three hawks marked in urban areas dispersed an average of 123 km (51.7 SE) to breeding sites in exurban areas. Finally, three hawks

Figure 9.3. Power companies in some cities are responsive and proactive in reducing electrocutions by installing protective covers, such as the one this Harris's hawk is perched on, over energized equipment. Other raptors, such as great horned owls and red-tailed hawks, also benefit from these conservation actions. Photograph courtesy of James F. Dwyer.

marked in one urban area dispersed an average of 153 km (10.7 SE) across the Sonoran desert to other urban areas where they were found dead.[7] Additional information on dispersal from areas other than Tucson would be informative in understanding connectivity within and between areas of breeding habitat.

DENSITY

Estimates for nest densities in exurban areas range from 2 to 5 nests per 10 km² in Arizona, fewer than 1 to 2 nests per 10 km² in New Mexico, and 2.3 nests per 10 km² in Texas.[4] The average density for two study areas in Arizona was 1 nest per 2.5 km², with average distances between nesting pairs of 1.8 km and a closest distance of 0.5 km.[15] In comparison, the number of nesting groups in urban Tucson was estimated to have increased from 10 in 1975, to 46 in 1993, to 62 in 2003,[4] but a lack of quantitative description of amount of area searched precluded density estimates. In the absence of information describing nest density throughout the rest of the species' range, these numbers lack the context necessary to be truly informative from broader conservation and management perspectives.

Community Ecology

INTERSPECIFIC INTERACTIONS

As is common among most large hawks, Harris's hawks are frequently mobbed by American kestrels (*Falco sparverius*) and several species of small songbirds such as western kingbirds (*Tyrannus verticalis*) and scissor-tailed flycatchers (*Tyrannus forficatus*). In urban areas, they are especially harassed by northern mockingbirds (*Mimus polyglottus*). Although undoubtedly annoying, these present no risk to the Harris's hawk.

Just as these smaller birds see Harris's hawks as a threat, Harris's hawks display aggression toward other species perceived as threats to them or their nests in both urban and exurban locations. These aggressive behaviors may be limited to perching near the threat and vocalizing alarm calls or may escalate to physical attacks in which the hawks fly at and strike the individual seen as a threat.

A study of the nesting habitat of five raptors in Tucson, Arizona, offers a glimpse into the urban community ecology of Harris's hawks. Both Harris's hawks and Cooper's hawks (*Accipiter cooperii*) most frequently built their nests in Aleppo pine (*Pinus halepensis*) and species of *Eucalyptus* trees but partitioned the urban environment when selecting nesting locations.[27] Harris's hawks primarily hunt rabbits and quail, which are present in Tucson but are more common in areas with natural vegetation. As expected, Harris's hawks typically selected nesting areas with a lower density of buildings and more interspersed native vegetation. In contrast, Cooper's hawks, which are primarily bird predators, nested in high-density residential areas dominated by introduced vegetation. Red-tailed hawks (*Buteo jamaicensis*) and burrowing owls (*Athene cunicularia*) generally used areas of the city that were largely undeveloped. Great horned owls capture a wide variety of prey but do not build their own nests. As would be expected, great horned owls were found relatively evenly distributed across the city of Tucson, where they took advantage of existing hawk nests, apparently regardless of where they had been built.[27]

INTERACTIONS WITH HUMANS

Harris's hawks were historically considered quite wary of human intrusions toward their nests in exurban areas,[5,9,28] but this may be changing in the urban setting. Responses of Harris's hawks in exurban areas include making alarm calls while soaring over the area and intruders, or leaving the nest area. In contrast, Harris' hawks in urban areas are incredibly habituated to human presence, often nesting unbothered in the yards of residences. Also, in comparison to some urban raptors, such as Mississippi kites (*Ictinia mississippiensis*)[29] and red-shouldered hawks (*Buteo lineatus*),[30] which may become aggressive toward human pedestrians, Harris's hawks almost never strike humans. In fact, the only reported aggressive act of an urban Harris's hawk toward humans involved an escaped falconry bird that had been captive-reared.[7] It appears that Harris's hawks are very compatible with humans in urban settings.

Conservation and Management

Conservation and management of exurban Harris's hawks focus primarily on reducing habitat loss and avoiding disturbance of family groups during nesting. Though urban-nesting individuals are typically well-habituated to the movement and noise of human residents, exurban birds are more accustomed to solitude. These individuals are easily disturbed, leaving their nests while people are present in the area and potentially causing the death of eggs or young from heat or cold, depending on conditions. Wild Harris's hawks, prized for their ability to hunt as part of a team with humans and dogs, also are still legally collected from the wild for use in falconry.

Conservation and management of Harris's hawks in urban areas primarily focus on mitigating electrocution risk, particularly near nests.[24] Harris's hawk family groups frequently perch together on a single structure,[4] and particularly in urban areas, the behavior often occurs on power poles. This can substantially increase the likelihood that at least one individual will perch in a dangerous location, resulting in an electric shock injury or an electrocution death.[24,25] To a lesser extent, urban Harris's hawks encounter many of the other urban risk factors experienced by other urban-dwelling raptors.[26]

Across exurban and urban environments, the majority of research on Harris's hawks has occurred north of the US-Mexico border, but the distribution of Harris's hawks is primarily south of that border. Our understanding of Harris's hawk biology, social ecology, urban adaptation, and conservation and management needs will be incomplete as long as this pattern of disproportionate study persists.

Literature Cited

1. Dawson, J. W. 1988. "The Cooperative Breeding System of the Harris' Hawk in Arizona." MS thesis, University of Arizona.

2. Ellis, D. H., J. C. Bednarz, D. G. Smith, and S. P. Flemming. 1993. "Social Foraging Classes in Raptorial Birds." *BioScience* 43:14–20.

3. Ferguson-Lees, J., and D. A. Christie. 2001. *Raptors of the World*. Boston: Houghton Mifflin.

4. Dwyer, J. F., and J. C. Bednarz. 2011. "Harris's Hawk (*Parabuteo unicinctus*)." In *The Birds of North America*, edited by P. G. Rodewald. Ithaca: Cornell Lab of Ornithology. Accessed February 22, 2017. https://birdsna.org/Species-Account/bna/species/hrshaw.

5. Mader, W. J. 1975. "Extra Adults at Harris's Hawk Nests." *Condor* 77:482–85.

6. Dawson, J. W., and R. W. Mannan. 1989. "A Comparison of Two Methods of Estimating Breeding Group Size in Harris' Hawks." *Auk* 106:480–83.

7. Dawson, J. W., and R. W. Mannan. 1994. "The Ecology of Harris' Hawks in Urban Environments. Arizona Game and Fish Department, Final Report." Urban Heritage Grant LOA G20058-A, Tucson, AZ.

8. Beingolea, O. 2010. "Harris' Hawks in Lima, Peru." *American Falconry* 54:28–48.

9. Bednarz, J. C. 1987. "Pair and Group Reproductive Success, Polyandry, and Cooperative Breeding in Harris's Hawks." *Auk* 104:393–404.

10. Sheehy, R. R. 1995. "A Phylogenetic Analysis of the Accipitridae (Class Aves)." PhD diss., University of Arizona.

11. Dawson, J. W., and R. W. Mannan. 1991a. "Dominance Hierarchies and Helper Contributions in Harris' Hawks." *Auk* 108:649–60.

12. Dawson, J. W., and R. W. Mannan. 1991b. "The Role of Territoriality in the Social Organization of Harris's Hawks." *Auk* 108:661–72.

13. Jiménez, J. E., and F. M. Jaksić. 1993. "Observations on the Comparative Behavioral Ecology of Harris' Hawk in Central Chile." *Journal of Raptor Research* 27:143–48.

14. Coulson, J. O., and T. D. Coulson. 2012. *The Harris's Hawk Revolution*. Pearl River, NY: Parabuteo Publishing.

15. Whaley, W. H. 1986. "Population Ecology of the Harris' Hawk in Arizona." *Raptor Research* 20:1–15.

16. Hayden, T. J., and J. C. Bednarz. 1991. "The Los Medaños Cooperative Raptor Research and Management Program, Final Report 1988–1990." University of New Mexico, Albuquerque, NM.

17. Brannon, J. D. 1980. "The Reproductive Ecology of a Texas Harris's Hawk (*Parabuteo unicinctus harrisi*) Population." MS thesis, University of Texas.

18. Pavez, E. F., G. A. Lobos, and F. M. Jaksić. 2010. "Long-Term Changes in Landscape and in Small Mammal and Raptor Assemblages in Central Chile." *Revista Chilena de Historia Natural* 83:99–111.

19. Bednarz, J. C. 1988. "Cooperative Hunting in Harris's Hawks (*Parabuteo unicinctus*)." *Science* 239:1525–27.

20. Bednarz, J. C., and J. D. Ligon. 1988. "A Study of the Ecological Bases of Cooperative Breeding in the Harris' Hawk." *Ecology* 69:1176–87.

21. Coulson, J. O., and T. D. Coulson. 2013. "Reexamining Cooperative Hunting in Harris's Hawk (*Parabuteo unicinctus*): Large Prey or Challenging Habitats?" *Auk* 130:548–52.

22. Hedlund. C. A. 1998. "*Trichomonas gallinae* in Avian Populations in Urban Tucson, Arizona." MS thesis, University of Arizona.

23. Coulson, J. O., and T. D. Coulson. 2003. "Harris's Hawk Dies after Colliding with Motor Vehicle." *Journal of Raptor Research* 37:350–51.

24. Dwyer, J. F., and R. W. Mannan. 2007. "Preventing Raptor Electrocutions in an Urban Environment." *Journal of Raptor Research* 41:259–67.

25. Dwyer, J. F. 2006. "Electric Shock Injuries in a Harris's Hawk Population." *Journal of Raptor Research* 40:193–99.

26. See chapter 14.

27. Mannan, R. W., C. W. Boal, W. J. Burroughs, J. W. Dawson, T. S. Estabrook, and W. S. Richardson. 2000. "Nest Sites of Five Raptor Species along an Urban Gradient." In *Raptors at Risk: Proceedings of the V World Conference on Birds of Prey and Owls*, edited by R. D. Chancellor and B.-U. Meyburg, 447–53. Berlin, Germany: World Working Group on Birds of Prey and Owls; Surrey, BC: Hancock House Publishers.

28. Whaley, W. H. 1979. "The Ecology and Status of the Harris's Hawk (*Parabuteo unicinctus*) in Arizona." MS thesis, University of Arizona.

29. See chapter 6.

30. See chapter 17.

CHAPTER 10

Barred Owls: A Nocturnal Generalist Thrives in Wooded, Suburban Habitats

Richard O. Bierregaard

IN AN OLD RESIDENTIAL NEIGHBORHOOD of Charlotte, North Carolina, the loud, baritone call of a barred owl (*Strix varia*) breaks the quiet of a late winter evening, not so much asking as demanding to know "Who cooks for you, who cooks for you-all?" His mate responds, in a slightly higher key, and their conversation escalates into a raucous caterwauling of bizarre hoots. A passerby could not be faulted for thinking there was a monkey up there in the trees somehow involved in the discussion. All across the southeastern United States in suburban areas with mature canopies of deciduous trees, this scene is repeated throughout the spring and early summer. Barred owls have taken to this suburban habitat like ducks to water.

Barred owls are nonmigratory, primarily nocturnal forest-dwelling birds of prey with a strong fondness for mature forests.[1,2] In eastern North America, they can be found through moist forests from Florida, continuing north to the Gaspe Peninsula in eastern Quebec and west to the edge of the prairies from eastern Texas all the way to southern Manitoba.[3] Beginning late in the 19th century and through the early 20th, this resourceful species extended its range west. They now make their homes in parts of Oregon, Washington, most of British Columbia, and southern Alaska and the Northwest Territories.[4,5]

In some areas, especially in the southeastern United States as far south as Florida, as north as Washington, DC (D. Johnson, pers. comm.), and as far west as Cincinnati, Ohio,[6] barred owls are firmly established in suburban neighborhoods old enough to have a canopy of mature hardwoods. In the northeastern United States, barred owls nest in suburban habitats in Westchester County, New York (Anne Swaim, pers. comm.), and southern Connecticut[7] but not at the notable densities seen in the southeast. In the Pacific northwest, barred owls nest in urban-suburban environments in the lower Fraser Valley in British Columbia[9] and are commonly found in parks and forest tracts in Seattle (E. Deal, pers. comm.) and adjacent Bainbridge Island in Washington,[8] but thus far they are not as intimately integrated into suburban neighborhoods as in the southeastern United States.

Despite their widespread distribution in urban areas, few studies of the species have been carried out in these environments. Dykstra et al.[6] described their habitat characteristics in and around Cincinnati, Ohio; Hindmarch and Elliott[9] described prey species used in urban environments in British Columbia; and my students and I studied the dense suburban barred owl population and neighboring rural populations in and around Charlotte, North Carolina (color plate 8). During the 11 years of our study (2001–11), we followed between 3 and 55 pairs each year, collecting data on nest-site characteristics and home range,[10] dispersal and source-sink ecology,[11] prey selection,[12] and mortality in environments with dense human populations.[13] These studies form the basis for much of the discussion in this chapter.

Behavior and Ecology

Habitat

Across their range, barred owls nest in an unsurprisingly wide range of forest types, from mixed coniferous forests in the northwest to lowland hardwood-bald cypress (*Taxodium distichum*) swamps in the southeast. Despite this diversity, these forests share common features associated with mature stands. Mature forests typically have a relatively closed canopy, with trees large enough to provide nest cavities and a relatively open understory, which facilitates the barred owls' sit-and-wait hunting style.[2] The trees of older (>50–60 years) wooded suburban neighborhoods in the southeast provide a sort of *uber*-mature forest habitat. Mature hardwoods provide ample cavities for nesting, and the understory of mowed lawns and ornamental shrubs and hedges, interspersed with parks and streamside vegetation, provides ideal hunting conditions.

Nest-Site Selection

In both suburban and rural areas, barred owls nest primarily in hardwood tree cavities, branch scars, or chimney-like snags.[10,14] Rarely, old stick nests of crows (*Corvus brachyrhynchos*) or hawks (typically red-shouldered hawks [*Buteo lineatus*]) are used.[6,14] They also readily accept nest boxes. In suburban Charlotte, barred owls are regularly rescued from chimneys they were presumably exploring as potential nest sites. We rescued one owl from two different chimneys several years apart, and one pair successfully raised young in a decorative (i.e., nonfunctional) chimney for two years before moving into a nest box we installed nearby.

Suburban-nesting barred owls typically used large nest trees (Charlotte, North Carolina, average diameter at breast height [DBH] = 95 cm, n = 36; Cincinnati, Ohio, mean DBH = 77 cm, n = 7).[6] Average DBH for rural nests in Charlotte was 63 cm (n = 16). Nest cavities ranged from less than 2 to 22 m above ground (mean = 10.3 m, n = 37), and cavity floor area in Charlotte averaged 0.763 cm^2 (n = 23). In Charlotte, rural nests were more often in dead snags than were suburban nests.[10] This may be due primarily to a lack of availability in urban regions rather than any preferences of the owl; dead and dying trees in suburban areas are often removed for human safety concerns.[10] In our suburban population, willow oaks (*Quercus phellos*) were most commonly used because they are the dominant tree in the suburban landscape. Sycamores (*Platanus occidentalis*) also often provided suitable cavities; half of the barred owl nests studied by Dykstra et al.[6] in Cincinnati were in sycamores. The understory around rural nests in Charlotte was much more dense than that around suburban nests.[10] Suburban and rural nest sites also are almost always close to water.[3] Our study area in Charlotte provides a network of streamside areas, and the dense human population does not deter owls from nesting.

Although some barred owl nests in suburban Charlotte were in parks and greenways, most were in yards in residential neighborhoods (figure 10.1). In both rural and suburban areas, vegetation and environmental measures around active nest trees and randomly selected plots did not differ, but there were significant differences in the habitats surrounding rural and suburban nests.[10] These suburban nests were, on average, only 11 m from a house and 41 m from the nearest road. In Cincinnati, 15 ha plots surrounding the nest sites averaged 48 percent forest and 31 percent residential development, with some nests within 20 m of residences.[6] In urban areas of British Columbia, owls nested in residential parks; the area surrounding the nest sites ranged from 54 percent to 90 percent urban (3 km^2 circular plots centered on the nest).[9]

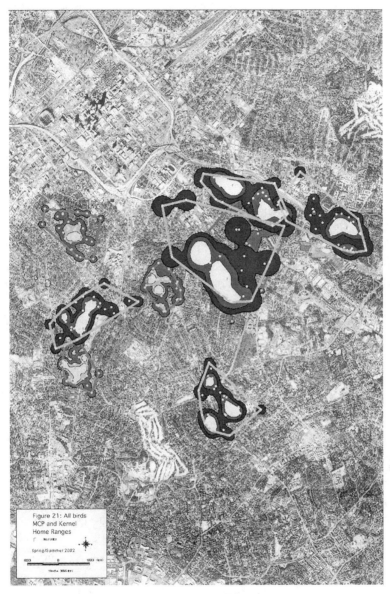

Figure 10.1. Minimum convex polygon and kernel home-range estimates for eight adult barred owls in suburban Charlotte, North Carolina. Larger black ranges are males, and smaller dark gray ranges are females. Photo courtesy of Eric Harrold.[10]

Although territories can be stable over decades, nests in natural cavities tend to be ephemeral. Damage to trees can leave former cavities unusable, and squirrels can fill cavities with their own nesting material, rendering them useless to the owls. One of our suburban pair of owls used four different cavities in a four year period.

Home Ranges

Excluding one owl pair with an abnormally large range, the average home-range size for male and female barred owls in suburban Charlotte was 86 and 44 ha, respectively (figure 10.1).[10] We documented 19 pairs in an approximately 17 km² part of our suburban study site, where we were confident that we had located all territories (density of 1.1 pairs/km²). Because of the difficulties in tracking owls in dense hardwood forests, we were unable to measure comparable home ranges from our rural populations. However, we documented four pairs of nesting owls in an area of approximately 150 ha in a rural nature preserve. Two of the pairs nested only 270 m apart. Although this density was higher than that found in our suburban study area, it would not be typical of rural densities at the landscape level. At this larger scale, suitable patches of forests exist within a matrix of open land inhospitable to the owls, so their density would be considerably lower.

In non-suburban habitats in the northern portion of the species' range, radio-tagged barred owls used home ranges of 149 to 321 ha.[15,16,17,18] These were all substantially larger than those recorded in either our suburban or rural study areas.

Diet

The barred owl is a generalist predator, taking a wide range of small- to medium-sized prey ranging from insects to earthworms, crawfish, mammals, reptiles, amphibians, and birds.[1] In the northern reaches of the species' range, small mammals predominate (>70 percent of prey items identified in 43 studies) in the diet.[1] The only study of barred owl diet in the southeastern part of their range was a comparison of suburban and rural owls in and around Charlotte, North Carolina.[12] There were dramatic differences in diet between the two populations, based on more than 1,500 prey items recorded in eight suburban and three rural nests over a three year period (table 10.1). Avian prey represented 54.5 percent of the deliveries to nests in the suburban population; all were taken at night. In

Table 10.1. Diet of rural and suburban populations of Charlotte, NC, versus other rural North American areas.[1]

Population	Mammals		Birds		Reptiles/ Amphibians		Fish		Crayfish		Insects		Total
	n	%	n	%	n	%	n	%	n	%	n	%	
Rural	76	15.2	27	5.4	156	31.3	9	1.8	85	17.0	146	29.3	499
Suburban	164	17.3	552	58.3	51	5.4	50	5.3	40	4.2	90	9.5	947
Livezey 2007	5085	71.9	669.8	9.5	469.1	6.6	133.7	1.9	173.2	2.4	546.2	7.7	7077

contrast, avian prey represented only 4.9 percent of prey items in rural nests, where reptiles and amphibians were the most frequently delivered prey items (28.3 percent). This disparity likely reflects the clear differences in understory vegetation between rural and suburban environments. In suburban areas, the near absence of any understory other than ornamental bushes and hedges in most of the owls' home ranges makes it much easier to see roosting birds and reduces the amount of habitat for reptiles, amphibians, and small mammals. In our study, barred owls delivered similar amounts of mammalian prey at the suburban (16.2 percent) and rural (13.8 percent) sites, but this was markedly less than in other published studies, most of which were conducted farther north. In northern areas of the barred owls' range, climate conditions prevent reptiles and amphibians from being available in the early part of the breeding cycle, so mammals represent a much higher percentage of the diet, typically over half of prey items delivered in the breeding season[20,21] and as high as 97 percent in winter.[22] In urban British Columbia, birds accounted for only 4.5 percent of prey items found in pellets at roosts and nest sites; nonnative rats (*Rattus norvegicus*) dominated the diet and were increasingly important as urban habitat surrounding the nest site increased.[9]

OTHER BEHAVIORS

Barred owls are almost exclusively monogamous. However, we found three adults attending a nest at one suburban site in our study area. Video recordings at the nest showed two females simultaneously—one atop the other—incubating three eggs in a branch-scar depression. On one of the recordings, a female perched on the rim of the nest cavity, with the second female incubating. When the male arrived with a mouse, the perched female took the mouse from him and passed it on to the female on the eggs. We do not know whether both birds contributed to the clutch (which did not hatch).

Contrary to Mazur and James's[3] statement that barred owls do not breed until they are at least one year old, we documented four cases of owls less than a year old (males and females) successfully breeding. In one case, the young male replaced an established (radio-tagged) male that had been killed in a collision with a car. Our data indicated that the lack of observations of barred owls breeding in their first year of life is likely not because they are physiologically incapable but more likely because most juveniles are unable to find an available territory or mate in their first year of life. In suburban Charlotte, with apparently high mortality rates (see the following), vacancies in established territories open frequently and are quickly filled from the population of unmated birds, which includes recently fledged young.

Population Ecology

Reproduction

In Charlotte, most eggs are laid in February, with no difference between rural and suburban birds. The average clutch size in Charlotte is typically just over two eggs (2.1, $n = 121$); only one nest had four eggs. This is lower than other reports for the species, which range from 2.7 in Wisconsin[19] to 2.3 in Nova Scotia,[20] and agrees with the trend for smaller clutch sizes at lower latitudes reported for barred owls east of the Appalachians.[23]

In both rural and suburban habitats, young leave the nest at 4–5 weeks of age and move about the branches of their nest tree. This relatively early departure from the nest is presumably an antipredator adaptation. Once they have left the nest cavity, they never return to it. By 5 weeks of age (not 10–12 weeks as reported earlier[24]), they are able to fly short distances from tree to tree. The young stay close to one another, often perching shoulder to shoulder. Presumably, this keeps the family together as young begin to explore and facilitates prey delivery by the adults, which continues up to two months after the young leave nests in Charlotte.

Reproductive success, measured as the number of young fledged per pair that laid eggs, of 46 suburban pairs was higher than that of 12 rural pairs (1.50 vs. 1.04 young per pair, respectively; 95 total nesting attempts over three years).[11] However, the pooled reproductive success of the suburban and rural Charlotte pairs (1.38) was lower than the rates (1.38–2.09) reported in three studies in the northern portion of the species' range.[14,19,20] The low reproductive success of the rural birds in our study may be the result of the relatively young age of the forests occupied by our rural population. Young forests have dense understories, which make hunting more difficult, and trees with nest cavities so small that young often fall from the nests. In one extreme example, one of the young fell from a nest cavity with a floor of only 18 cm in diameter.

Survival

The longevity record for barred owls is from an owl recovered dead 26 years after it was banded (USGS Bird Banding Laboratory, unpublished data). Barred owls in the Charlotte suburbs seem to have a relatively high mortality rate, primarily as a result of collisions with vehicles.[13] Indeed, barred owls are the most frequently admitted species at the Carolina Raptor Center rehabilitation facility, and 47.9 percent of those admissions result from vehicular collisions. Additionally, although territories were consistently occupied over the course of our study, we documented a high turnover of individual breeders in the suburban population—again, often as a result of vehicular mortality. One territory that has been occupied by owls for decades according to local residents had three females and two males during the 11 years of our study. At another nest site, three males sequentially used the territory during our study, but the female remained throughout. With this apparently high mortality rate and turnover, it is reasonable to question whether the suburban area constitutes a source or sink (area that produces a net gain or net loss in numbers, respectively) for the population.[25,26] However, the stable population in our suburban area, short distances covered by dispersing young (see the following), and the rapid replacement of adults lost suggest that this is not an ecological sink.[11]

Immigration, Emigration, and Dispersal

In one study, we radio-tagged 8 young owls in rural and 14 young owls in suburban populations in and around Charlotte, North Carolina, to compare juvenile dispersal in the two environments.[11] We followed the birds until they settled down, usually within a few months of leaving their natal ranges, and discovered that young in rural nests dispersed farther (4.2 km) than those from suburban nests (2.7 km). The owls in our rural study areas nested in patches of woods surrounded by fields. When the young started to disperse, they had to cross these open areas to get to the nearest forest. If they found a neighboring pair already in residence, the dispersing juveniles would have to push on to the next remnant of forest, and so on. In contrast, the suburban owls in large part found hospitable habitat surrounding them, so they could find an unoccupied neighborhood in which to settle closer to their natal range than the rural young.

Community Ecology

Intra- and Interspecific Interactions

Barred owls and red-shouldered hawks are considered diurnal and nocturnal ecological equivalents,[27] which is to say that they live in the same habitat and prey on mostly the same or similar species, but the barred owls work the night shift and the hawks work the day shift. The urban and suburban nesting habitats of the two species in Cincinnati are virtually indistinguishable,[6] and prey species are similar as well.[12,28] In suburban Charlotte, a pair of nests used by the two species were less than 50 m apart. Although there are often some vocal interactions as the hawks are settling in at the end of the day and the owls are beginning their nocturnal activity, as a rule, a peaceful coexistence is established.

However, our observations revealed that barred owls will prey on other raptors, including screech-owls (*Megascops* spp.) and sharp-shinned hawks (*Accipiter striatus*), and in turn fall prey to great horned owls (*Bubo virginianus*).[29] The absence of barred owls in some suburban areas correlates with the presence of great horned owls (e.g., Ithaca, New York [K. McGowan, pers. comm.], the Philadelphia suburbs [R. Bierregaard, unpubl. data], and Baton Rouge, Louisiana [P. Stouffer, pers. comm.]) but not in Cincinnati (C. Dykstra, pers. comm.). It is not clear whether the presence of great horned owls inhibits barred owls or if the habitat is simply more suitable for great horned owls. In both Ithaca and Baton Rouge, barred owls are found in appropriate rural habitat surrounding the suburban neighborhoods that are occupied primarily by great horned owls. The rural area surrounding the older Philadelphia suburbs is a mosaic of forest and open space more suitable for great horned owls than barred owls, which may explain the relative scarcity of barred owls there, despite the environmental similarities to the land occupied so successfully by barred owls in Charlotte. It has been suggested the presence of barred owls in the Seattle, Washington, area is related to the proximity (<20 km) of extensive forests.[30]

INTERACTIONS WITH HUMANS

Barred owls in suburban environments have little fear of humans and can be aggressive near their nests, particularly when young inadvertently fall from their nest cavity or, in the case of recently branched young, wind up on the ground after a clumsy landing. Using their beaks and talons, these young typically can climb back up all but the smoothest-barked trees. Regardless, while the young are on the ground, their parents may attack humans that approach them.

There is a rather surprising increase of owl attacks on humans in late summer and early fall in suburban habitats near Charlotte. This occurs at a time of the year when adult territoriality is waning and young are dispersing from their natal areas, so there would seem to be no motivation for adult owls to attack humans. One possible explanation is that these are young owls playfully attacking moving targets. This behavior has been reported by a rehabilitator who had two orphaned young flying at liberty around his house. He witnessed the young owls sallying out to strike people passing on foot or on bicycles in what can only be described as play.

Conservation and Management

Globally, the species is certainly secure. Somewhat paradoxically, given the species' adaptability to nesting in human-dominated suburban habitats, it is classified as a species of concern in New Jersey,[31] where it has been negatively affected by human-dominated landscapes.[32] In addition to collision with vehicles, the use of second-generation anticoagulant rodenticides may pose a threat to suburban populations of barred owls and is worth monitoring.[9,33] In the Pacific Northwest, the species has expanded into the range of the endangered northern spotted owl (*Strix occidentalis caurina*) and has depressed spotted owl numbers through usurpation of their territories and hybridization.[34] Experimental programs of lethal removal of barred owls from spotted owl ranges are ongoing.[35,36]

As newer suburban neighborhoods begin to develop a forest canopy, barred owls from adjacent productive populations may colonize these areas before trees are old enough to have cavities large enough for successful breeding, presumably creating an ecological sink. In these situations, nest boxes are an ideal management approach. Well-constructed nest boxes (see Barker and Wolfson[37] for design plans) can bridge the gap until the canopy trees mature to the point where they can provide natural cavities.

Acknowledgments

I would like to thank the many people who contributed to the success of the research in Charlotte: graduate students Jennifer Bates, Cori Cauble, Eric Harrold, and Jim Mason spent hundreds of hours in the field tracking radio-tagged owls and pouring over video recordings from our nest cameras as they collected data for their theses; a small flock of UNC–Charlotte undergraduates helped with the field and lab work; dozens of landowners graciously allowed us to install nest boxes and traipse through their yards at night; and the Carolina Raptor Center provided logistical support and access to their database covering four decades of barred owl admissions. Clint Boal, Cheryl Dykstra, and Courtney Lix greatly improved the chapter with their keen editorial input.

Literature Cited

1. Livezey, K. B. 2007. "Barred Owl Habitat and Prey: A Review and Synthesis of the Literature." *Journal of Raptor Research* 41:177–201.

2. Bent, A. C. 1938. "Part 2: Orders Falconiformes and Strigiformes." *Life Histories of North American Birds of Prey.* Bulletin of the US National Museum, No. 170. Washington, DC.

3. Mazur, K. M., and P. C. James. 2000. "Barred Owl (*Strix varia*)." In *The Birds of North America,* edited by P. G. Rodewald. Ithaca: Cornell Lab of Ornithology. Accessed January 7, 2017. https://birdsna.org/Species-Account/bna/species/brdowl.

4. Livezey, K. B. 2009. "Range Expansion of Barred Owls, Part I: Chronology and Distribution." *American Midland Naturalist* 161:49–56.

5. Livezey, K. B. 2009. "Range Expansion of Barred Owls, Part II: Facilitating Ecological Changes." *American Midland Naturalist* 161:323–49.

6. Dykstra, C. R., M. M. Simon, F. B. Daniel, and J. L. Hays. 2012. "Habitats of Suburban Barred Owls (*Strix varia*) and Red-Shouldered Hawks (*Buteo lineatus*) in Southwestern Ohio." *Journal of Raptor Research* 46:190–200.

7. Yannielli, L. C. 1991. "Preferred Habitat of Northern Barred Owls in Litchfield County, Connecticut." *Connecticut Warbler* 11:12–20.

8. Acker, J. 2012. "Recent Trends in Western Screech-Owl and Barred Owl Abundances on Bainbridge Island, Washington." *Northwestern Naturalist* 93:133–37.

9. Hindmarch, S., and J. E. Elliott. 2015. "When Owls Go to Town: The Diet of Urban Barred Owls." *Journal of Raptor Research* 49:66–74.

10. Harrold, E. S. 2003. "Barred Owl (*Strix varia*) Nesting Ecology in the Southern Piedmont of North Carolina." MS thesis, University of North Carolina–Charlotte.

11. Mason, J. S. 2004. "The Reproductive Success, Survival, and Natal Dispersal of Barred Owls (*Strix varia*) in Rural versus Urban Habitats in and around Charlotte, NC." MS thesis, University of North Carolina–Charlotte.

12. Cauble, L. C. 2008. "The Diets of Rural and Suburban Barred Owls (*Strix varia*) in Mecklenburg County, North Carolina." MS thesis, University of North Carolina–Charlotte.

13. Gagné, S. A., J. L. Bates, and R. O. Bierregaard. 2015. "The Effects of Road and Landscape Characteristics on the Likelihood of a Barred Owl (*Strix varia*)-Vehicle Collision." *Urban Ecosystems* 18:1007–20.

14. Postupalsky, S., J. M. Papp, and L. Scheller. 1997. "Nest Sites and Reproductive Success of Barred Owls (*Strix varia*) in Michigan." In *Biology and Conservation of Owls of the Northern Hemisphere,* edited by J. R. Duncan, D. H. Johnson, and T. H. Nicholls, 325–37. Second International Symposium. Gen. Tech. Rep. NC-190. St. Paul, MN: USDA Forest Service, North Central Forest Experiment Station.

15. Mazur, K. M., S. D. Frith, and P. C. James. 1998. "Barred Owl Home Range and Habitat Selection in the Boreal Forest of Central Saskatchewan." *Auk* 115:746–54.

16. Nicholls, T. H., and M. R. Fuller. 1987. "Territorial Aspects of Barred Owl Home Range and Behavior in Minnesota." In *Biology and Conservation of Northern Forest Owls*, edited by R. W. Nero, R. J. Clark, R. J. Knapton, and R. H. Hamre, 121–28. Gen. Tech. Rep. RM-142. Fort Collins, CO: USDA Forest Service.

17. Elody, B., and N. Sloan. 1985. "Movements and Habitat Use of Barred Owls in the Huron Mountains of Marquette County, Michigan, as Determined by Radiotelemetry." *Jack-Pine Warbler* 63:3–8.

18. Hamer, T. E. 1988. "Home Range Size of the Northern Barred Owl and Northern Spotted Owl in Western Washington." MS thesis, Western Washington University.

19. Johnson, D. H. 1987. "Barred Owls and Nest Boxes—Results of a Five-Year Study in Minnesota." In *Biology and Conservation of Northern Forest Owls*, edited by R. W. Nero, R. J. Clark, R. J. Knapton, and R. H. Hamre, 129–34. Gen. Tech. Rep. RM-142. Fort Collins: USDA Forest Service.

20. Elderkin, M. F. 1987. "The Breeding and Feeding Ecology of a Barred Owl, *Strix varia* Barton, Population in Kings County, Nova Scotia." MS thesis, Acadia University.

21. Bosakowski, T., and D. G. Smith. 1992. "Comparative Diets of Sympatric Nesting Raptors in the Eastern Deciduous Forest Biome." *Canadian Journal of Zoology* 70:984–92.

22. Marks, J. S., D. P. Hendricks, and V. S. Marks. 1984. "Winter Food Habits of Barred Owls in Western Montana." *Murrelet* 65:28–29.

23. Murray, G. A. 1976. "Geographic Variation in the Clutch Size of Seven Owl Species." *Auk* 93:602–13.

24. Soucy, L. J., Jr. 1976. "Barred Owl Nest." *North American Bird Bander* 1:68–69.

25. Pulliam, H. R. 1988. "Sources, Sinks, and Population Regulation." *American Naturalist* 132:652–61.

26. Pulliam, H. R. 1996. "Sources and Sinks: Empirical Evidence and Population Consequences." In *Population Dynamics in Ecological Space and Time*, edited by J. O. E. Rhodes, R. K. Chesser, and M. H. Smith, 45–69. Chicago: University of Chicago Press.

27. Temple, S. A., and B. L. Temple. 1976. "Avian Population Trends in Central New York, 1935–72." *Journal of Field Ornithology* 47:238–57.

28. Dykstra, C. R., J. L. Hays, and S. C. Crocoll. 2008. "Red-Shouldered Hawk (*Buteo lineatus*)." In *The Birds of North America*, edited by P. G. Rodewald. Ithaca: Cornell Lab of Ornithology. Accessed January 7, 2017. https://birdsna.org/Species-Account/bna/species/reshaw.

29. Gerard, M., J. B. Clark, K. Kozij, and D. Zazelenchuk. 2009. "A Sample of Prey Remains Found In Great-Horned Owl Nests In Saskatchewan In 2008." *Blue Jay* 67:71–76.

30. Rullman, S., and J. M. Marzluff. 2014. "Raptor Presence along an Urban–Wildland Gradient: Influences of Prey Abundance and Land Cover." *Journal of Raptor Research* 48:257–72.

31. New Jersey Department of Environmental Protection. 2017. "Threatened and Endangered Species." Accessed April 6, 2017. http://www.nj.gov/dep/fgw/tandespp .htm.

32. Bosakowski, T., and D. G. Smith. 1997. "Distribution and Species Richness of a Forest Raptor Community in Relation to Urbanization." *Journal of Raptor Research* 31:26–33.

33. Albert, C. A., L. K. Wilson, P. Mineau, S. Trudeau, and J. E. Elliott. 2010. "Anticoagulant Rodenticides in Three Owl Species from Western Canada, 1988–2003." *Archives of Environmental Contamination and Toxicology* 58:451–59.

34. Gutiérrez, R. J., M. Cody, S. Courtney, and A. B. Franklin. 2007. "The Invasion of Barred Owls and Its Potential Effect on the Spotted Owl: A Conservation Conundrum." *Biological Invasions* 9:181–96.

35. Buchanan, J. B., R. J. Gutiérrez, R. G. Anthony, T. Cullinan, L. V. Diller, E. D. Forsman, and A. B. Franklin. 2007. "A Synopsis of Suggested Approaches to Address Potential Competitive Interactions between Barred Owls (*Strix varia*) and Spotted Owls (*S. occidentalis*)." *Biological Invasions* 9:679–91.

36. Diller, L. V., K. A. Hamm, D. A. Early, D. W. Lamphear, K. M. Dugger, C. B. Yackulic, C. J. Schwarz, P. C. Carlson, and T. L. McDonald. 2016. "Demographic Response of Northern Spotted Owls to Barred Owl Removal." *Journal of Wildlife Management* 80:691–707.

37. Barker, M. A., and E. Wolfson. 2013. *Audubon Birdhouse Book: Building, Placing, and Maintaining Great Homes for Great Birds*. Minneapolis: Voyageur Press.

CHAPTER 11

Powerful Owls:
Possum Assassins Move into Town

Raylene Cooke, Fiona Hogan, Bronwyn Isaac,

Marian Weaving, and John G. White

ONCE THOUGHT TO LIVE ONLY in large forested areas, the powerful owl (*Ninox strenua*), Australia's largest and most iconic of owls (figure 11.1), surprisingly is now turning up frequently in the cities of eastern Australia. Powerful owls require ample prey and large tree cavities for nest sites; how this top-order predator is able to survive in human-dominated landscapes is an important question for conservation and the focus of ongoing research. The powerful owl is endemic to Australia, resident in the three eastern mainland states and the Australian Capital Territory, and classified nationally as "rare."[2,3] First described by Gould in 1838, powerful owls are an unusual raptor in that they do not exhibit reversed sexual size dimorphism, the prevalent trait among raptors in which females are larger than males. For reasons still not understood, male powerful owls grow to a height of 65 cm and weigh up to 1,700 g, compared to females, which grow to a height of 54 cm and weigh up to 1,308 g.[1]

The powerful owl was once considered a specialist in ecological terms due to its restricted habitat and dietary requirements,[4,5,6] indicating that it was vulnerable to habitat modification and had specific conservation needs. Studies

Figure 11.1. Pair of powerful owls. Photo by John G. White.

also suggested that powerful owls were confined to forests more than 60 years old[7] and that they were dependent on arboreal marsupial prey that were themselves dependent on trees with hollows.[4,8] Loss of essential habitat, combined with the relatively small population size (approximately 7,000 breeding adults)[9] and restricted distribution,[10] has made the powerful owl a species of conservation interest both nationally[9] and internationally (Appendix II CITES and IUCN [2012 IUCN Red List of Threatened Species]).

Despite being a forest-dependant species, the powerful owl is present in highly fragmented urban and urban-fringe environments of major cities along the eastern seaboard.[11,12,13] Its presence in these areas calls into question the degree to which the powerful owl is vulnerable to habitat modification and the extent to which it has specific conservation needs. Much recent research, therefore, has concentrated on powerful owls in urban and urban-fringe areas, particularly in Melbourne, where in some instances these urban and suburban birds are successfully breeding.[14]

Undertaking Research on Powerful Owls

Research on powerful owls has used primarily noninvasive techniques, such as surveys and observations, to focus on diet, roost and nest tree characteristics, and breeding success. Much of the information on these ecological attributes comes from owls in natural forested landscapes and rural areas, although researchers have begun measuring these variables in urban environments, including Melbourne, Brisbane, and Sydney. With urban areas expanding and powerful owls increasingly utilizing these landscapes, it is critical to understand what drives the occurrence and survival of owls in urban landscapes.

Diet

Powerful owls are opportunistic, nocturnal predators that prey predominately on medium to large (100–1,200 g) arboreal, hollow-dependent marsupials[15] such as possums. In keeping with being opportunistic hunters, however, their diet composition varies spatially and temporally.[7,15,16,17,18] Powerful owl diet in urban areas is somewhat consistent with their diet in forested areas, in that medium-sized arboreal marsupials are the primary food source. With common ringtail possums (*Pseudocheirus peregrinus*) and common brushtail possums (*Trichosurus vulpecula*) abundant throughout urban areas in eastern Australia, these two species predominate the diet of powerful owls in urban areas (color plate 9).[15,19]

Roosting Sites

During the breeding season, the female powerful owl remains in the nest hollow, while the male roosts in a regular roost tree usually within 50 m of the nest hollow.[20] Outside the breeding season, both adult birds spend their day roosting in selected trees. Roost tree characteristics are consistent between urban and forested environments.[21] Roost trees have dense crowns, which suggests a requirement for seclusion, and although the owls use trees of varying heights, they roost only in the top one-third of the tree.[21] Favored roost trees include various *Eucalyptus* species, blackwoods (*Acacia melanoxylon*), other wattles (*Acacia* spp.), rainforest trees, tall casuarinas (*Casuarina littoralis*), and native cherry (*Exocarpos cupressiformis*).[8,22]

Nest Trees

Nest trees used by the owls to rear young are typically *Eucalyptus* species with large cavities for breeding.[12] Suitable hollows do not develop in trees until they are aged 150 to 500 years old.[23,24] Thus losses of large old trees due to intensive timber harvesting[25] and urbanization threaten the persistence of the powerful owl, as they lead to a reduction of suitable hollows for nesting and/or for the owl's prey species.[26]

Breeding

The powerful owl is a long-lived species with a relatively low reproductive rate.[17] It is a seasonal breeder[17] with only one nesting attempt per year and a maximum of two fledglings per attempt.[8] Pairs that fail during their initial breeding attempt will abandon the nest without any further attempts until the next year.[4,17,27]

Powerful owls will breed in urban areas; however, a lack of suitable breeding hollows[19,28] and increased disturbance levels[29] are their greatest limiting factors. Where suitable hollows are present in urban areas, powerful owls have bred successfully, with young fledging over consecutive years in parklands less than 20 km from Melbourne's central business district. When pairs have suitable resources in terms of prey, roosting habitat, and nest cavities, there appears to be no difference in the breeding success rates of urban- and forest-dwelling owls.

Determining Presence

Given the cryptic, nocturnal nature of powerful owls, determining their presence can be problematic, and the perceived rarity of the owl could be largely a result of low observer effort and inappropriate survey techniques.[11] Owl sightings or evidence of presence (e.g., vocalizations, regurgitated pellets, or excreta) provide a straightforward measure to accurately detect presence; however, determining presence without these cues can be challenging.

The most widely used technique to determine presence of powerful owls uses call playback. Call playback is a survey method that takes advantage of a target species' territorial behavior, in which it will respond when it hears an intruder by calling or flying toward the surveyor. The method involves the audio broadcast of a recorded vocalization of the target species, after which the surveyor listens and watches for a response. Although the use of playback has proved successful with many raptor species, including spotted owls (*Strix occidentalis*),[30] screech-owls (*Megascops asio*),[31] and tawny owls (*Strix aluco*),[32] playback as a detection method for powerful owls has proved problematic. A minimum of 18 site visits is required to be confident of a powerful owl absence,[33] and both season and temperature influence the response rate.[34] Surveys were more successful in eliciting responses from owls in summer and spring, with nightly temperatures greater than 20°C.[34]

Incorporating Modern Technologies

The last 20 years have seen an expansion of research on powerful owls, with continuing studies examining ecological traits such as breeding, habitat use, and diet. Building on this growing body of research, new technologies now facilitate much greater insights into powerful owl ecology in Australia. Several techniques in particular can help address the challenges associated with finding and tracking owls, including spatial modeling, GPS tracking, and DNA profiling. These methodological approaches are improving our understanding of this species' ecology, with implications for studying other urban raptors as well.

Spatial Modeling

The inability to detect cryptic species across large spatial scales or gradients produces a substantial impediment to conservation. Advances in presence-only modeling and the current availability of presence datasets have provided tools to predict potential occurrence and investigate how changes in environments or resources affect species. The program MaxEnt (version 3.3.3)[35] determines the spatial probability distribution of a species across a defined area, based on known species presences and eco-geographical variables (EGVs).[35,36] To understand the effect of increasing urbanization on powerful owls in southeastern Australia, we employed MaxEnt presence-only modeling followed by Multi-Criteria Decision Analysis (MCDA). MCDA is a form of site suitability analysis within a Geographic Information Systems (GIS) platform used to weight resource layers based on their importance in order to refine habitat suitability models.[37,38,39]

Our initial modeling focused on the powerful owl,[28] with later modeling assessing potential prey and tree cavity occurrence to produce layers for MCDA.[14,19,40] We collected presence data for the models from fieldwork and supplemented these where possible with citizen science atlas datasets. We employed measures to restrict spatial and historical biases associated with presence data derived from atlas datasets.[28] We selected or derived eco-geographical variables based on the ecology of the owl or prey species. Models incorporated both default and alternate settings where ecologically applicable. Initial modeling for the powerful owl indicated the importance of vegetated environments, particularly those in close proximity to riparian systems, for the occurrence of this species.[28] Additional analysis showed that forests and urban-fringe environments are capable of supporting enough potential habitat for powerful owls to successfully forage, roost, and nest, but urban areas of the gradient provide only limited and highly scattered patches of potential habitat for this species. This suggests a decline in potential habitat as urbanization intensifies.[28]

Modeling the suitable habitat for a target species, based on occurrence records and EGVs alone, delineates all the potential habitats a species may occupy. This fails to incorporate factors, such as prey occurrence or reproductive resources, that limit whether a patch is actually suitable or not for a given species. To account for these important spatially explicit resources for the powerful owl (e.g., prey and tree cavities suitable for nesting), we completed additional presence-only modeling on these resources,[19,40] using general prey, diverse prey, and large cavity layers to constrain our initial owl model under realistic resource scenarios.[14] This allowed us to delineate which habitat might be suitable for owls to settle (i.e., areas with suitable prey) and which might be suitable for successful reproduction (i.e., areas with suitable prey and trees large enough for cavities). The proportional difference between suitable habitat for settlement versus suitable habitat for reproduction could be seen as the potential for the powerful owl to make a maladaptive habitat choice.

Relative to increasing urbanization, powerful owls are likely to make maladaptive habitat choices during dispersal because, although urban environments provide plentiful prey, they generally lack tree cavities large enough for this species to breed. Through modeling and MCDA, we have identified the potential occurrence of the powerful owl along an urban-to-forest gradient, underscoring the possibility for this species to make a maladaptive habitat choice with increasing urbanization. This illustrates the potential for urban environments to form an ecological trap due to the lack of suitable tree cavities for breeding.[14]

GPS TRACKING

Although modeling is useful as a conservation tool, if we want to clearly understand the biology, behavior, and ecology of a species, we also need to employ methods such as tracking and genetic profiling. There are three main methods for tracking raptors: Very High Frequency radio-tracking (VHF), global positioning system (GPS) tracking, and satellite tracking. The method used depends on the research questions and the target species. For investigation of raptor migration, satellite transmitters are the best option, although they are the most expensive, because data are sent to satellites and relayed to a ground station for direct download. GPS transmitters are generally better for research on localized home range, habitat use, and resource partitioning. GPS trackers store data on the transmitter itself, and for the earliest models, the animal needed to be recaptured to access the collected data. New advances in GPS trackers, such as solar cells to extend battery life and remote download capabilities, have revolutionized GPS tracking. These advances have provided the opportunity to investigate more thoroughly the spatial movements and habitat use of powerful owls.

Prior to the advent of GPS trackers, powerful owls in forested habitat in southeastern New South Wales[41] and in the box-ironbark-eucalypt woodlands of central Victoria[42] were tracked with VHF transmitters. Until recently, GPS tracking was not feasible because the technology was not suitable for our purposes and because capturing this species was very difficult. However, in 2016, we attached GPS trackers for the first time to five urban powerful owls (three females and two males; figure 11.2) in the period leading up to breeding (February to June), and these trackers have yielded 5,149 individual data points. Home ranges for these owls ranged from 388 ha to 1,805 ha, with four of the five urban home ranges smaller than 870 ha (female average = 460 ha; male average = 1,334 ha). This was considerably smaller than the average home ranges reported for owls living in natural, forested areas (female average = 1,773 ha; male average = 2,333 ha).[43]

Figure 11.2. Researcher releasing a powerful owl with a GPS tracker attached to the tail. Photo by John G. White.

Combining GPS Tracking with Spatial Modeling

Given that we had developed predictive models for powerful owl habitat suitability across urban Melbourne and now had GPS data on their actual movements, the next logical step was to combine these to verify our habitat suitability model. When these data were combined, it became clear that urban powerful owls are primarily using these predicted areas as their core home ranges and only deviating into unsuitable areas occasionally. We also tested the alignment of powerful owl home ranges with protected areas (e.g., national parks, state parks, local government reserves). In some cases, much of the habitat used by the urban owls was not within the protected area network. These nonprotected areas, although providing habitat now, are at risk due to urban expansion and intensification. This potentially has important long-term implications for these owls; as the residential properties become more urban due to increased urban sprawl, these owls may lose a substantial amount of habitat within their home ranges to urban development. The combined use of species distribution modeling and validation with actual tracking data provides strong support for focusing urban planning on preservation of predicted suitable areas and reducing the degree to which urbanization is allowed to intensify.

GENETIC PROFILING

Individual identification is one of the most useful yet challenging aspects of ecological studies. Although previous breeding studies have been undertaken on the powerful owl[29,44] and many owlets have been banded over the decades,[44] positive individual identification is difficult, as feathers often obscure bands on the tarsus. Genetic profiling can be used to identify individuals by using a suite of microsatellite markers together with a sex-linked marker to generate a genetic tag.[45] Once obtained, genetic tags can be used to investigate mating systems, dispersal, and landscape genetics. Genetic tags can be developed from DNA collected invasively (by taking blood, performing tissue biopsy, or plucking feathers) or noninvasively (by collecting shed feathers).[46]

Investigating mating systems using genetic profiling

Genetic profiling of shed feathers offers unprecedented potential for studying the mating systems of the powerful owl. Because powerful owls occupy regular roosts,[20,21] shed feathers from individuals can be collected from beneath roosting owls, eliminating the need for catching or even observing the owl. Powerful owls are notoriously difficult to catch; thus, such noninvasive methods offer a great opportunity. DNA can be obtained from shed feathers collected from territorial adults at the start of the breeding season and from blood samples taken from nestlings prior to fledging. Genetic profiles generated can then be used to investigate extra-pair fertilization, as well as mate fidelity (i.e., Do the same individuals pair together over consecutive years?), nest fidelity (i.e., Does the same pair use the same nest over consecutive years?), and relatedness of the breeding pair.

Genetic profiling of shed feathers collected from breeding powerful owls from 2003 to 2005 provided the first insights into the mating systems of the powerful owl.[45] We studied pairs of owls located in the urban fringe ($n = 4$ pairs) and, as a comparison, in the continuous forest habitat ($n = 3$ pairs). Mate fidelity was high among breeding pairs, with no evidence of extra-pair fertilization in either habitat. Pairs occupied a single nest site over consecutive years, and no partners were displaced during the study. However, we detected evidence of inbreeding in two pairs of owls in the urban-fringe habitat, in which one pair of owls were half-siblings and the other pair were full siblings. From a conservation standpoint, inbreeding is a concern as it can result in deleterious traits in offspring and decrease the biological fitness of a population. It also suggests that inbreeding avoidance mechanisms may be nullified due to the lack of potential mates, which may be a result of low local recruitment into the population. Opportunistic rather than selective mating in powerful owls is therefore an increasing concern for small urban populations.

Investigating dispersal using genetic profiling

Genetic tags can be used to identify an individual owl throughout its life, allowing researchers to learn its dispersal and/or fate in one of two ways. In the first case, the genetic tag of a nestling is matched to its own feathers or other samples collected later in life. In the second, an individual is matched to its parents through parent-pair analysis.[47] Using these methods, we identified five offspring that were all progeny of an urban-fringe pair that occupied the same nest site for 10 years.[45] Two of the individuals identified were deceased fledglings found close to the nest site, while the other three offspring were located in the urban fringe, 2.5 km to 18 km from the natal territory.[45] Most samples that were used in the parent-pair analysis in this instance were shed feathers collected opportunistically by citizen scientists.

Next Steps

The presence of powerful owls in major urban centers across eastern Australia is important, both for the species and for public appreciation of urban biodiversity, as with this appreciation comes increased public pressure for protection. The powerful owl is iconic and therefore inspires much public interest in its protection. Although this species is surviving in urban environments, we believe these environments are also acting as an ecological trap for many pairs, as the resources (specifically, nest cavities) required for breeding are generally not present.[14] Although only a single event, the recent successful breeding of a powerful owl pair in a nest box in an urban area provides hope for the potential use of nest boxes as a conservation management strategy to alleviate the limited cavity resources for powerful owls in urban environments.[48] New technologies such as GPS tracking, spatial modeling, and genetics have certainly improved our understanding of this species. There are, however, still many basic ecological questions about these urban birds that we cannot answer. One obvious gap in knowledge relates to juvenile dispersal in urban areas: we know little about where these juveniles go and whether they successfully pair and breed. As GPS tracking devices become smaller, lighter, and more efficient, attaching these to juvenile powerful owls may become possible in the not too distant future.

Continued urban development and urban sprawl is the single biggest threat to urban powerful owls. Therefore, it is critical that future land-use planning incorporates the protection and maintenance of suitable habitat, especially the retention of old-growth trees with suitable breeding hollows, to ensure retention of key habitat features for powerful owls.

Literature Cited

1. Olsen, P. D. 1991. "Do Large Males Have Small Testes? A Note on Allometric Variation and Sexual Size Dimorphism in Raptors." *Oikos* 60:134–36.
2. Garnett, S. 1992. "Threatened and Extinct Birds of Australia." Report No. 82. Melbourne, Australia: Royal Australasian Ornithologists Union.
3. Olsen, P. D. 1998. "Australia's Raptors: Diurnal Birds of Prey." Birds Australia Conservation Statement No. 2 (Supplement). Melbourne, Australia: Birds Australia.
4. Fleay, D. 1968. *Nightwatchmen of Bush and Plain*. Brisbane, Australia: Jacaranda Press.
5. Seebeck, J. H. 1976. "The Diet of the Powerful Owl *Ninox strenua* in Western Victoria." *Emu* 76:167–70.
6. Roberts, G. J. 1977. "Birds and Conservation in Queensland." *Sunbird* 8:73–82.
7. Kavanagh, R. P. 1988. "The Impact of Predation by the Powerful Owl *Ninox strenua* on a Population of the Greater Glider *Petauroides volans*." *Australian Journal of Ecology* 13:445–50.
8. Schodde, R., and I. J. Mason. 1980. *Nocturnal Birds of Australia*. Melbourne, Australia: Lansdowne.
9. Garnett, S. T., and G. M. Crowley. 2000. "The Action Plan for Australian Birds." *Environment Australia*, Canberra, Australia.
10. Higgins, P. J. E. 1999. *Handbook of Australian, New Zealand and Antarctic Birds, Vol. 4: Parrots to Dollarbird*. Melbourne, Australia: Oxford University Press.
11. Pavey, C. R. 1993. "The Distribution and Conservation Status of the Powerful Owl *Ninox strenua* in Queensland." In *Australian Raptor Studies*, edited by P. Olsen, 144–54. Melbourne, Australia: Australasian Raptor Association, Royal Australasian Ornithologists Union.
12. Cooke, R., R. Wallis, and A. Webster. 2002a. "Urbanization and the Ecology of Powerful Owls *(Ninox strenua)* in Outer Melbourne, Victoria." In *Ecology and Conservation of Owls*, edited by I. Newton, R. Kavanagh, J. Olsen, and I. Taylor, 100–106. Melbourne, Australia: CSIRO Publishing.
13. Kavanagh, R. P. 2004. "Conserving Owls in Sydney's Urban Bushland: Current Status and Requirements." In *Urban Wildlife: More than Meets the Eye*, edited by D. Lunney and S. Burgin, 93–108. New South Wales, Australia: Royal Zoological Society of New South Wales, Mosman.
14. Isaac, B., R. Cooke, D. Ierodiaconou, and J. White. 2014a. "Does Urbanization Have the Potential to Create an Ecological Trap for Powerful Owls (*Ninox strenua*)?" *Biological Conservation* 176:1–11.

15. Cooke, R., R. Wallis, F. Hogan, J. White, and A. Webster. 2006. "The Diet of Powerful Owls (*Ninox strenua*) and Prey Availability in a Continuum of Habitats from Disturbed Urban Fringe to Protected Forest Environments in South-Eastern Australia." *Wildlife Research* 33:199–206.

16. Cooke, R., R. Wallis, A. Webster, and J. Wilson. 1997. "Diet of a Family of Powerful Owls (*Ninox strenua*) from Warrandyte, Victoria." *Proceedings of the Royal Society of Victoria* 109:1–6.

17. Debus, S., and C. Chafer. 1994. "The Powerful Owl *Ninox strenua* in New South Wales." *Australian Birds* 28:S21–64.

18. Pavey, C. R. 1994. "Records of the Food of the Powerful Owl *Ninox strenua* from Queensland." *Sunbird* 24:30–39.

19. Isaac, B., J. White, D. Ierodiaconou, and R. Cooke. 2014b. "Simplification of Arboreal Marsupial Assemblages in Response to Increasing Urbanization." *PLoS ONE* 9: e91049.

20. Traill, B. J. 1993. "The Diet and Movement of a Pair of Powerful Owls *Ninox strenua* in Dry Forest." In *Australian Raptor Studies*, edited by P. Olsen, 155–69. Melbourne, Australia: Australasian Raptor Association, Royal Australasian Ornithologists Union.

21. Cooke, R., R. Wallis, and J. White. 2002b. Use of Vegetative Structure by Powerful Owls in Outer Urban Melbourne, Victoria, Australia-Implications for Management. *Journal of Raptor Research* 36:294–99.

22. Bilney, R. J., R. Cooke, and J. G. White. 2011. "Potential Competition between Two Top-Order Predators following a Dramatic Contraction in the Diversity of Their Prey Base." *Animal Biology* 61:29–47.

23. Ambrose, G. 1982. "An Ecological and Behavioural Study of Vertebrates Using Hollows in Eucalypt Branches." PhD thesis, La Trobe University.

24. Parnaby, H. 1995. "Hollow Arguments." *Nature Australia* 25:80.

25. Davey, S. M. 1993. "Notes on the Habitat of Four Australian Owl Species." In *Australian Raptor Studies*, edited by P. Olsen, 126–42. Melbourne, Australia: Australasian Raptor Association, Royal Australasian Ornithologists Union.

26. Isaac, B., R. Cooke, D. Simmons, and F. Hogan. 2008. "Predictive Mapping of Powerful Owl (*Ninox strenua*) Breeding Sites Using Geographical Information Systems (GIS) in Urban Melbourne, Australia." *Landscape and Urban Planning* 84:212–18.

27. Hollands, D. 1991. *Birds of the Night*. Sydney, Australia: Reed Books.

28. Isaac, B., J. White, D. Ierodiaconou, and R. Cooke. 2013. "Response of a Cryptic Apex Predator to a Complete Urban to Forest Gradient." *Wildlife Research* 40:427–36.

29. Webster, A., R. Cooke, G. Jameson, and R. Wallis. 1999. "Diet, Roosts and Breeding of Powerful Owls *Ninox strenua* in a Disturbed, Urban Environment: A Case for Cannibalism? Or a Case of Infanticide?" *Emu* 99:80–83.

30. Rinkevich, S. E., and R. J. Gutiérrez. 1966. "Mexican Spotted Owl Habitat Characteristics in Zion National Park." *Journal of Raptor Research* 30:74–78.

31. Carpenter, T. W. 1987. "Effects of Environmental Variables on Responses of Eastern Screech Owls to Playback." In *Biology and Conservation of Northern Forest Owls*, edited by R. W. Nero, R. J. Clark, R. J. Knapton, and R. H. Hamre, 121–28. Symposium proceedings. Gen. Tech. Rep. RM-142. Fort Collins, CO: USDA Forest Service, Rocky Mountain Forest and Range Experiment Station.

32. Redpath, S. M. 1994. "Censusing Tawny Owls *Strix aluco* by the use of imitation Calls." *Bird Study* 41:192–98.

33. Wintle, B. A., R. P. Kavanagh, M. A. McCarthy, and M. A. Burgman. 2005. "Estimating and Dealing with Detectability in Occupancy Surveys for Forest Owls and Arboreal Marsupials." *Journal of Wildlife Management* 69:905–17.

34. Cooke, R., H. Grant, I. Ebsworth, A. R. Rendall, and J. G. White. 2017. "Can Owls Be Used to Monitor the Impacts of Urbanisation? A Cautionary Tale of Variable Detection." *Wildlife Research* 44:573–81.

35. Phillips, S. J., R. P. Anderson, and R. E. Schapire. 2006. "Maximum Entropy Modeling of Species Geographic Distributions." *Ecological Modelling* 190:231–59.

36. Elith, J., S. J. Phillips, T. Hastie, M. Dudík, Y. E. Chee, and C. J. Yates. 2011. "A Statistical Explanation of MaxEnt for Ecologists." *Diversity and Distributions* 17:43–57.

37. Gurnell, J., M. J. Clark, P. W. W. Lurz, M. D. F. Shirley, and S. P. Rushton. 2002. "Conserving Red Squirrels (*Sciurus vulgaris*): Mapping and Forecasting Habitat Suitability Using a Geographic Information Systems Approach." *Biological Conservation* 105:53–64.

38. Powell, M., A. Accad, and A. Shapcott. 2005. "Geographic Information System (GIS) Predictions of Past, Present Habitat Distribution and Areas for Re-Introduction of the Endangered Subtropical Rainforest Shrub *Triunia robusta* (Proteaceae) from South-East Queensland Australia." *Biological Conservation* 123:165–75.

39. Sener, B., M. L. Süzen., and V. Doyuran. 2006. "Landfill Site Selection by Using Geographic Information Systems." *Environmental Geology* 49:376–88.

40. Isaac, B., J. White, D. Ierodiaconou, and R. Cooke. 2014c. "Urban to Forest Gradients: Suitability for Hollow Bearing Trees and Implications for Obligate Hollow Nesters." *Austral Ecology* 39:963–72.

41. Kavanagh, R. 1997. "Ecology and Management of Large Forest Owls in South-Eastern Australia." PhD diss., University of Sydney.

42. Soderquist, T., and D. Gibbons. 2007. "Home-Range of the Powerful Owl (*Ninox strenua*) in Dry Sclerophyll Forest." *Emu* 107:177–84.

43. Bradsworth, N., J. G. White, B. Isaac, and R. Cooke. 2017. "Species Distribution Models Derived from Citizen Science Data Predict the Fine Scale Movements of Owls in an Urbanizing Landscape." *Biological Conservation* 213:27–35.

44. McNabb, E. G., R. P. Kavanagh, and S. Craig. 2007. "Further Observations on the Breeding Biology of the Powerful Owl *Ninox strenua* in South-Eastern Australia." *Corella* 31:6–9.

45. Hogan, F. E., and R. Cooke. 2010. "Insights into the Breeding Behaviour and Dispersal of the Powerful Owl (*Ninox strenua*) through the Collection of Shed Feathers." *Emu* 110:178–84.

46. Hogan, F., R. Cooke, C. Burridge, and J. Norman. 2008. "Optimizing the Use of Shed Feathers for Genetic Analysis." *Molecular Ecology Resources* 8:561–67.

47. Kalinowski, S. T., M. L. Taper, and T. C. Marshall. 2007. "Revising How the Computer Program CERVUS Accommodates Genotyping Error Increases Success in Paternity Assignment." *Molecular Ecology* 16:1099–106.

48. McNabb, E., and J. Greenwood. 2011. "A Powerful Owl Disperses into Town and Uses an Artificial Nest-Box." *Australian Field Ornithology* 28:65–75.

CHAPTER 12

Burrowing Owls:
Happy Urbanite or Disgruntled Tenant?

Courtney J. Conway

BURROWING OWLS (*ATHENE CUNICULARIA*) OFTEN stand on a fence post or on top of a mound of dirt, bobbing their heads, looking at you with large yellow eyes, and standing tall on their long legs (color plate 10). Affectionately known as the "howdy owl," burrowing owls are unique because they nest and roost in underground burrows (figure 12.1). Burrowing owls inhabit urban and suburban areas in North and South America[1,2,3,4,5,6] and are often promoted by wildlife agencies as charismatic and harmless members of the local "watchable wildlife" community. They are generally popular and appreciated by nature enthusiasts. In much of their range, burrowing owls do not dig their burrows but rather depend on other burrowing animals (e.g., prairie dogs or ground squirrels) to provide the burrows that they call home. But burrowing owls will also use a variety of holes and crevasses in human-made structures. Adults use the burrows for protection from predators, for shade from the hot sun, and as a nest where they raise their young. Females lay up to 11 eggs in an underground chamber of the burrow. When the eggs hatch, the young owls stay safe below ground for several weeks. Adults defend their burrow and their chicks from predators and from intrusion by other burrowing owls. Nestlings also scare off

Figure 12.1. Adult burrowing owl standing next to the entrance to the owl's underground nest burrow in southeastern Washington.

predators by imitating a rattlesnake rattle,[7] which helps deter would-be predators. Because they spend so much time deep underground in their burrows where carbon dioxide levels can be high, burrowing owls have adaptations to help them survive in a high carbon dioxide environment.[8] It is safe to say that the life of the burrowing owl revolves around its burrow.

Burrowing owls nesting in urban areas often rely on small patches of semi-natural open space within suburban or industrial areas.[1,9,10] In particular, burrowing owls frequent open areas such as airports, golf courses, parks, school yards, landfills, cemeteries, university campuses, industrial compounds, irrigation canals, culverts, roadway berms, military facilities, vacant lots within residential areas, and even backyards of homes (figure 12.2).[4,9,11,12,13,14] As a result, some researchers have said that burrowing owls have the ability to coexist or even thrive with humans in urban settings or have suggested that they are less affected by human development than other raptors.[15,16] In contrast, other researchers have suggested or demonstrated that urbanization, at least in some contexts, can harm burrowing owl populations.[4,9] Indeed, many burrowing owl

Figure 12.2. A burrowing owl nesting in a rock pile within a residential area in south-eastern Washington.

population declines appear to be caused by urbanization.[4,12,13,17,18] For example, breeding populations have nearly disappeared completely from highly urban areas near San Francisco, Bakersfield, Davis, and San Diego, California.[19,20] Moreover, many other possible causes of population declines in burrowing owl populations are human related, especially reductions in numbers of burrowing mammals, use of pesticides, and conversion of native grassland to agriculture or development.[5,13,14,17,21,22,23]

In one sense, one might expect burrowing owls to be more susceptible to the impacts of urbanization than other raptors because they spend a considerable amount of time standing on the ground near nest burrows, they require deep holes in the ground for nesting, and their nesting burrows are likely a limiting resource.[24,25,26,27,28] Indeed, numerous studies have reported that human development (urban, rural, and industrial) is an important threat to the species.[13,18,29,30] But in another sense, burrowing owls may seem less susceptible to urbanization and human disturbance because, behaviorally, they can be quite tolerant of human activity near their nests.[31] Furthermore, the relationship between burrowing owls and urbanization is not straightforward and undoubtedly depends on land use, type of urbanization, density and type of buildings, amount of open space, prey availability, burrow availability, fate of burrowing mammal populations, and other factors that vary among cities.

As an area becomes more urbanized, burrowing owls living in the region may persist by coping with the urbanization. However, owls may also colonize an urban area that is attractive to them due to the new conditions or resources that the urban development creates (e.g., airports, golf courses). Conditions that attract burrowing owls to urban areas may include (1) increased abundance of burrowing mammals compared to the surrounding rural areas;[23] (2) loose dirt from construction and roadway maintenance (including road berms), which may allow burrowing mammals to dig more burrows in urban areas;[13] (3) increased prey abundance (insects and small rodents) compared to adjacent rural areas due to irrigation and other alterations;[32] (4) improved nighttime foraging opportunities due to street or security lights that attract and concentrate insect prey;[13,29,33] or (5) the creation of open areas suitable for burrowing owls in locations where tree or shrub density was previously too high. Urban areas in more arid portions of the owl's breeding range may have higher prey abundance than surrounding rural areas due to the increased moisture and higher biomass of green vegetation associated with urbanization.[6,32,34] These changes may attract burrowing owls, but they may produce an ecological trap if survival or reproduction is lower in the urban areas.

The effects of urbanization on burrowing owls seem to vary throughout the owl's range based on regional variation in the types of burrows that the owls use.[35] For example, the distribution and abundance of black-tailed prairie dogs (*Cynomys ludovicianus*), yellow-bellied marmots (*Marmota flaviventris*), and American badgers (*Taxidea taxus*) appear to be negatively associated with urban areas, but other species on which burrowing owls depend—such as round-tailed ground squirrels (*Xerospermophilus tereticaudus*), California ground squirrels (*Otospermophilus beecheyi*), and banner-tailed kangaroo rats (*Dipodomys spectabilis*)—seem to be more compatible with urban development, at least to some extent. In addition, burrowing owls more often use urban areas in the portions of their breeding range where they do not rely on prairie dogs, marmots, or badgers for nest burrows.[35] Specifically, most of the urban areas that support (or once supported) nesting burrowing owls are outside of the distribution of prairie dogs: Arizona (Tucson, Phoenix, Yuma), Texas (Corpus Christi), California (San Diego, Davis, Sacramento, Fremont/Newark, Oakland, Bakersfield, San Jose), Idaho (Boise), Washington (Pasco, Richland, Kennewick, Moses Lake), Florida (Cape Coral, Palm Beach/Boynton Beach, Pompano Beach/Ft. Lauderdale, Punta Gorda, Miami, Marco Island). Hence the effects of urbanization on burrowing owls depend on the specific nature of land uses and density of development and their influence on the vegetation, prey, burrows, and burrow providers. In some cases, urbanization benefits owls, while in others it is detrimental.

Behavioral Ecology

NEST-SITE SELECTION

Burrowing owls in urban areas nest in a variety of sites and contexts. Nesting burrowing owls require open areas with short grass and available burrows. Vegetation near the nest burrow is typically less than six inches tall and there must be adequate foraging habitat nearby to supply the invertebrates, mammals, and other small animals that constitute the diet of the burrowing owl. In western North America, owls in urban (and rural) areas typically nest in burrows created by burrowing mammals, such as prairie dogs and ground squirrels. The owls often use burrows in road and railway berms,[30] levees and ditches, and in disturbed open areas such as airport grasslands, golf courses, and fallow or grazed land.[10,36] Urban burrowing owls nesting in New Mexico were more likely to reuse nest burrows than those in rural areas, presumably because fewer suitable burrows were available in urban areas.[32] They also occasionally nest under concrete sidewalks and in drain pipes, piles of discarded tires, piles of riprap, or hay bales.[29] Burrowing owls seem to preferentially nest near roads, at least in some portions of their range.[22,23,37,38]

DIET

Burrowing owls eat a wide range of small animals and take advantage of locally abundant prey.[39,40] Their generalist diet is undoubtedly one of the traits that allows them to nest in a variety of urban settings. In some urban areas, burrowing owls forage under streetlights and seem to preferentially forage and perch near roads in portions of their range.[41,42] Although insects and spiders are typically the most numerous types of prey items, rodents often account for the bulk of the biomass of the birds' diet.[6,40] The diet of burrowing owls is similar at rural and urban sites in Florida, but owls at urban sites eat more avian prey compared to their rural counterparts.[6] Owls in New Mexico have higher foraging efficiency (presumably due to higher prey abundance) in urban areas, which compensates for greater human disturbance.[32] In contrast, hunting success, provisioning rates, and activity budgets are similar between urban and rural sites in Texas, but owls eat more flying insects at urban than at rural sites.[43] In an urban area in California, invertebrates are the most frequent prey throughout the year, but rodents contribute most to the biomass of the owls' diet in both breeding and nonbreeding seasons.[40]

OTHER BEHAVIORS

Owls readily take advantage of infrastructures in the urban environment by perching on objects such as fence posts, telephone wires, walls, and buildings.[29] Burrowing owls also nest in artificial burrows and some researchers and managers have installed artificial burrows in urban and suburban areas throughout the burrowing owl's range.[10,44,45,46,47] Artificial burrows in urban areas often serve as a mitigation and conservation tool for burrowing owls, and such efforts may delay the owl's extirpation from areas undergoing development.

Another behavioral characteristic that suits burrowing owls in urban settings is their ability to become quite habituated to humans. Burrowing owls are more wary of humans in rural than in urban areas.[26,27] When approached by a human, burrowing owls in rural areas fly off much sooner (i.e., when humans were farther away) than those in urban areas.[48] Individual burrowing owls seem to vary in their inherent fear of humans, and owls that are less fearful may be more prone to live in urban areas.[2] These differences might come about because owls raised in urban nests may be more conditioned to human presence and activities or because owls with a higher tolerance of human activities may preferentially settle in urban environments while those that lack tolerance settle in rural environments.[48]

An interesting behavior of burrowing owls is that, throughout their range, they collect mammalian manure (primarily cow or horse manure) and scatter it in and around the entrance to their nest burrows (figure 12.3).[41,49,50] In urban areas, they commonly use alternate materials in the same fashion, including paper, plastic, cotton, wood, dried vegetables, aluminum foil, cigarette butts, and shreds of carpet.[38,50,51] The reason burrowing owls engage in this odd behavior has been a topic of much debate and inquiry. Some scientists thought the owls did this to hide the scent of their offspring down in the burrow, but more recent research suggests that its purpose may be to entice insects to approach their nest burrows so the offspring can eat without having to leave the safety of the burrow.[50,51,52]

Figure 12.3. Mammal manure around the entrance of a burrowing owl nest burrow that was only 1 m from a busy road in an urban area in Tucson, Arizona.

Population Ecology

REPRODUCTION

The potential influences of urbanization on the nesting success and reproduction of burrowing owls are not clear. For example, the effect of urbanization on reproductive parameters in Florida burrowing owls was nonlinear and differed among reproductive measures. The amount of urban development in Florida apparently does not affect nesting success (defined as the percentage of nests where at least one young fledged), but the proportion of nests that fail from human-related causes grows with increasing development.[9] The number of young owls produced per nest site in Florida is positively correlated with development at sites with less than 60 percent development. However, the number of young produced per successful nest is negatively correlated with development in areas with more than 60 percent development.[9] In New Mexico, overall nesting success is lower, but the number of young produced per successful nest is higher in urban areas compared to native grasslands.[32] Another study in New Mexico reported that, within the urban environment, nesting success is higher for nests in areas with human-altered conditions compared to those in more natural areas.[53] In Texas, the number of young produced per owl pair is similar among nests in agricultural areas, industrial urbanized areas, or residential urbanized areas.[54] In eastern Washington, burrowing owls in an urban area have higher natal recruitment and adult return rates, lower nesting density and nesting success, and similar clutch size and nestlings per successful nest compared to nests in an agricultural area.[5] In Argentina, breeding success is higher in urban areas compared to rural areas.[48] Overall, reproduction metrics appear highly variable among locations for this species.[38] This may have more to do with environmental conditions associated with latitudinal and longitudinal gradients than specific local anthropogenic features or activities. More research is needed regarding the relationship between urbanization and nesting success (and other demographic traits) in burrowing owls.

SURVIVAL

In terms of survival, the comparative advantage or disadvantage of occupying urban areas is not clear. Mortality factors may be higher in urban areas compared to natural sites or at least perceived as such by burrowing owls.[53] For example, burrowing owls in urban areas in Texas are more vigilant than those in rural areas despite the lower numbers of potential owl predators observed near urban nest sites.[43] Owls nesting in urban areas in Texas experience more human disturbance compared to those in rural areas.[43] Additionally, burrowing owls frequently use roadways and culverts, which makes them susceptible to being hit and killed by vehicles.[37,38] Indeed, vehicle collisions often kill burrowing owls in many urban areas,[9,11,32,34,41] and vehicle collisions are more common during late summer and early fall when juveniles are dispersing from their natal areas.[55] However, survival rates in Argentina are higher for burrowing owls nesting in urban areas compared to those in rural areas, presumably because there are fewer predators in urban areas.[48]

DENSITY

Burrowing owl populations have declined throughout the species' range in western North America, and their breeding range has also contracted.[21,23,56,57,58] Unfortunately, we have no information regarding whether immigration, emigration, or dispersal differs between owls that nest in urban areas and those in rural areas or how this may influence persistence across the species distribution. Burrowing owl breeding densities seem to be higher in urban areas compared to nearby rural areas in many locations,[2,3,5] but there are exceptions.[32] In Florida, the density of burrowing owl nests is positively correlated with the percentage of development of low to moderate housing densities but then becomes negatively correlated once 45–60 percent of lots are developed.[9] The number of nesting pairs is similar among agricultural areas, industrial urbanized areas, and residential areas in Texas.[54] In both urban and non-urban areas, burrowing owls have an affinity for roads and culverts and other similar human-made structures. Wintering density of burrowing owls in Texas is higher in areas with more culverts and roads,[59] and the number of irrigation and roadside ditches or banks best predicts the abundance of nesting burrowing owls in a large agricultural landscape in California.[60] In Canada, burrowing owls are also associated with roads, but their use of roads decreases as average vehicle speed increases.[61]

Conservation and Management

Urban and suburban development may initially increase habitat suitability for burrowing owls by converting areas that were formerly unsuitable into nesting habitat (e.g., by clearing trees and shrubs, loosening soil to promote digging, and attracting insects and small mammals). But as the development and urbanization become more intensive, the owls often vacate these areas. Burrowing owls that nest on lots where home construction is occurring produced more young if a 10 m buffer from disturbance is provided around the nest burrow.[34] The presence of adequate foraging habitat is essential for burrowing owl survival. Burrowing owls tend to do the majority of their foraging within 550 m of their nest burrows. Burrowing owls in a California agricultural area have home ranges that vary from 93 to 140 ha.

Burrows are another critical resource for western burrowing owls because the owls rarely dig burrows themselves; they depend on the abandoned burrows of burrowing mammals (though the Florida subspecies does excavate its own burrows).[28] Burrowing owls have relatively high site and burrow fidelity from year to year, so persistence of the burrowing mammals that create the burrows likely influences population trends and persistence of burrowing owls. Hence, one key to the persistence of western burrowing owls at urban sites is the long-term presence of burrowing mammals within urban areas. Control efforts to eradicate burrowing animals are common in some urban areas frequented by burrowing owls, such as golf courses, airports, cemeteries, school yards, and residential lots.[10] Indeed, managers and researchers have removed or translocated burrowing owls from many airports due to the perceived risk of airstrikes,[10] and owls have also been moved to avoid conflict with urban development.[15] Some municipalities and state agencies have explicit policies and protocols for translocating burrowing owls out of areas slated for development.[62] The effectiveness of these actions has not been adequately documented, and there is no evidence that they are benefiting the species. Carefully controlled, replicated research studies to evaluate whether these efforts are beneficial or harmful to the viability of local burrowing owl populations have yet to be conducted; such studies would help ensure that mitigation policies are supported by science.

Acknowledgments

Any use of trade, firm, or product names is for descriptive purposes only and does not imply endorsement by the US government. J. Barclay, L. Trulio, and C. Lix provided helpful comments and suggestions that improved the chapter.

Literature Cited

1. Millsap, B. A. 2002. "Survival of Florida Burrowing Owls along an Urban-Development Gradient." *Journal of Raptor Research* 36:3–10.
2. Carrete, M., and J. L. Tella. 2011. "Inter-Individual Variability in Fear of Humans and Relative Brain Size of the Species Are Related to Contemporary Urban Invasion in Birds." *PLoS ONE* 6: e18859.
3. Rodríguez-Martínez, S., M. Carrete, S. Roques, N. Rebolo-Ifrán, and J. L. Tella. 2014. "High Urban Breeding Densities Do Not Disrupt Genetic Monogamy in a Bird Species." *PLoS ONE* 9: e91314.
4. Trulio, L. A., and D. A. Chromczak. 2007. "Burrowing Owl Nesting Success at Urban and Parkland Sites in Northern California." In *Bird Populations Monograph No. 1*, edited by J. H. Barclay, K. W. Hunting, J. L. Lincer, J. Linthicum, and T. A. Roberts, 115–22. Proceedings of the California Burrowing Owl Symposium, November 2003. Point Reyes Station: The Institute for Bird Populations.
5. Conway, C. J., V. Garcia, M. D. Smith, L. A. Ellis, and J. Whitney. 2006. "Comparative Demography of Burrowing Owls within Agricultural and Urban Landscapes in Southeastern Washington." *Journal of Field Ornithology* 77:280–90.
6. Mrykalo, R. J., M. M. Grigione, and R. J. Sarno. 2009. "A Comparison of Available Prey and Diet of Florida Burrowing Owls in Urban and Rural Environments: A First Study." *Condor* 111:556–59.
7. Owings, D. H., M. P. Rowe, and A. A. Rundus. 2002. "The Rattling Sound of Rattlesnakes (*Crotalus viridis*) as a Communicative Resource for Ground Squirrels (*Spermophilus beecheyi*) and Burrowing Owls (*Athene cunicularia*)." *Journal of Comparative Psychology* 116:197–205.

8. Kilgore, D. L., Jr., F. M. Faraci, and M. R. Fedde. 1985. "Ventilatory and Intrapulmonary Chemoreceptor Sensitivity to Carbon Dioxide in the Burrowing Owl (*Athene cunicularia*)." *Respiration Physiology* 62:325–40.

9. Millsap, B. A., and C. Bear. 2000. "Density and Reproduction of Burrowing Owls along an Urban Development Gradient." *Journal of Wildlife Management* 64:33–41.

10. Barclay, J. H. 2007. "Burrowing Owl Management at Mineta San Jose International Airport." In *Bird Populations Monograph No. 1*, edited by J. H. Barclay, K. W. Hunting, J. L. Lincer, J. Linthicum, and T. A. Roberts, 146–54. Proceedings of the California Burrowing Owl Symposium, November 2003. Point Reyes Station, CA: The Institute for Bird Populations.

11. Mealey, B. 1997. "Reproductive Ecology of the Burrowing Owls, *Speotyto cunicularia floridana*, in Dade and Broward Counties, Florida." In *The Burrowing Owl, Its Biology and Management, Including the Proceedings of the First International Symposium*, edited by J. L. Lincer and K. Steenhof. *Journal of Raptor Research*. Report 9:74–79.

12. Arrowood, P. C., C. A. Finley, and B. C. Thompson. 2001. "Analyses of burrowing owl populations in New Mexico." *Journal of Raptor Research* 35:362–70.

13. Klute, D. S., L. W. Ayers, M. T. Green, W. H. Howe, S. L. Jones, J. A. Shaffer, S. R. Sheffield, and T. S. Zimmerman. 2003. "Status Assessment and Conservation Plan for the Western Burrowing Owl in the United States." USDI Fish and Wildlife Service, Biological Technical Publication FWS/BTP-R6001–2003, Washington, DC.

14. Poulin, R. G., L. D. Todd, E. A. Haug, B. A. Millsap, and M. S. Martell. 2011. "Burrowing Owl (*Athene cunicularia*)." In *The Birds of North America*, edited by P. G. Rodewald. Ithaca: Cornell Lab of Ornithology. Accessed February 17, 2017. https://birdsna.org/Species-Account/bna/species/burowl.

15. Delevoryas, P. 1997. "Relocation of Burrowing Owls during Courtship Period." In *The Burrowing Owl, Its Biology and Management, Including the Proceedings of the First International Symposium*, edited by J. L. Lincer and K. Steenhof. *Journal of Raptor Research*. Report 9:138–44.

16. Martin, D. J. 1973. "Selected Aspects of Burrowing Owl Ecology and Behavior." *Condor* 75:446–56.

17. Grinnell, J., and A. H. Miller. 1944. "Distribution of the Birds of California." *Pacific Coast Avifauna* 27:202–3.

18. Konrad, P. M., and D. S. Gilmer. 1984. "Observations on the Nesting Ecology of Burrowing Owls in Central North Dakota." *Prairie Naturalist* 16:129–30.

19. Unitt, P. 2004. *San Diego County Bird Atlas*. San Diego Society of Natural History. El Cajon, CA: Sunbelt Publications.

20. Wilkerson, R. L., and R. B. Siegel. 2010. "Assessing Changes in the Distribution and Abundance of Burrowing Owls in California, 1993–2007." *Bird Populations* 10:1–36.

21. Desmond, M. J., J. E. Savidge, and K. M. Eskridge. 2000. "Correlations between Burrowing Owl and Black-Tailed Prairie Dog Declines: A 7-Year Analysis." *Journal of Wildlife Management* 64:1067–75.

22. Belthoff, J. R., and R. King. 2002. "Nest-Site Characteristics of Burrowing Owls (*Athene cunicularia*) in the Snake River Birds of Prey National Conservation Area, Idaho, and Applications to Artificial Burrow Installation." *Western North American Naturalist* 62:112–19.

23. Conway, C. J., and K. L. Pardieck. 2006. "Population Trajectory of Burrowing Owls in Eastern Washington." *Northwest Science* 80:292–97.

24. Thomsen, L. 1971. "Behavior and Ecology of Burrowing Owls on the Oakland Municipal Airport." *Condor* 73:177–92.

25. Haug, E. A., and A. B. Didiuk. 1993. "Use of Recorded Calls to Detect Burrowing Owls." *Journal of Field Ornithology* 64:188–94.

26. Conway, C. J., and J. Simon. 2003. "Comparison of Detection Probability Associated with Burrowing Owl Survey Methods." *Journal of Wildlife Management* 67:501–11.

27. Conway, C. J., V. Garcia, M. D. Smith, and K. Hughes. 2008. "Factors Affecting Detection of Burrowing Owl Nests during Standardized Surveys." *Journal of Wildlife Management* 72:688–96.

28. Coulombe, H. N. 1971. "Behavior and Population Ecology of the Burrowing Owl, *Speotyto cunicularia*, in the Imperial Valley of California." *Condor* 73:162–76.

29. Botelho, E. S., and P. C. Arrowood. 1998. "The Effect of Burrow Site Use on the Reproductive Success of a Partially Migratory Population of Western Burrowing Owls (*Speotyto cunicularia hypugaea*)." *Journal of Raptor Research* 32:233–40.

30. Dechant, J. A., M. L. Sondreal, D. H. Johnson, L. D. Igl, C. M. Goldade, P. A. Rabie, and B. R. Euliss. 1999. *Effects of Management Practices on Grassland Birds: Burrowing Owl.* Jamestown, ND: US Geological Survey Northern Prairie Wildlife Research Center.

31. COSEWIC. 2006. "COSEWIC Assessment and Update Status Report on the Burrowing Owl *Athene cunicularia* in Canada." Committee on the Status of Endangered Wildlife in Canada. Accessed February 17, 2017. http://www.registrelep-sararegistry .gc.ca/virtual_sara/files/cosewic/sr_burrowing_owl_e.pdf.

32. Berardelli, D., M. J. Desmond, and L. Murray. 2010. "Reproductive Success of Burrowing Owls in Urban and Grassland Habitats in Southern New Mexico." *Wilson Journal of Ornithology* 122:51–59.

33. Estabrook, T. S. 1999. "Burrow Selection by Burrowing Owls in an Urban Environment." MS thesis, University of Arizona.

34. Wesemann, T. W., and M. Rowe. 1987. "Factors Influencing the Distribution and Abundance of Burrowing Owls in Cape Coral, Florida." In *Integrating Man and Nature in the Metropolitan Environment: Proceedings of a National Symposium on Urban Wildlife*, edited by L. W. Adams and D. L. Leedy, 129–37. Columbia, MD: National Institute for Urban Wildlife.

35. Conway, C. J. 2018. "Spatial and Temporal Patterns in Population Trends and Burrow Usage of Burrowing Owls in North America." *Journal of Raptor Research.*

36. Trulio, L. A. 1997. "Burrowing Owl Demography and Habitat Use at Two Urban Sites in Santa Clara County, California." In *The Burrowing Owl, Its Biology and Management, Including the Proceedings of the First International Symposium,* edited by J. L. Lincer and K. Steenhof. *Journal of Raptor Research.* Report 9:84–89.

37. Plumpton, D. L. 1992. "Aspects of Nest-Site Selection and Habitat Use by Burrowing Owls at the Rocky Mountain Arsenal, Colorado." MS thesis, Texas Tech University.

38. Haug, E. A., B. A. Millsap, and M. S. Martell. 1993. "Burrowing Owl (*Speotyto cunicularia*)." In *The Birds of North America,* no. 61, edited by A. Poole and F. Gill. Philadelphia: The Academy of Natural Sciences; Washington, DC: American Ornithologists' Union.

39. Poulin, R. G., and L. D. Todd. 2006. "Sex and Nest Stage Differences in the Circadian Foraging Behaviors of Nesting Burrowing Owls." *Condor* 108:856–64.

40. Trulio, L. A., and P. Higgins. 2012. "The Diet of Western Burrowing Owls in an Urban Landscape." *Western North American Naturalist* 72:348–56.

41. Scott, T. G. 1940. "The Western Burrowing Owl In Clay County, Iowa, In 1938." *American Midland Naturalist* 24:585–93.

42. Marsh, A., E. M. Bayne, and T. I. Wellicome. 2014. "Using Vertebrate Prey Capture Locations to Identify Cover Type Selection Patterns of Nocturnally Foraging Burrowing Owls." *Ecological Applications* 24:950–59.

43. Chipman, E. D. 2006. "Behavioral Ecology of Western Burrowing Owls (*Athene cunicularia hypugaea*) in Northwestern Texas." MS thesis, Texas Tech University.

44. Trulio, L. A. 1995. "Passive Relocation: A Method to Preserve Burrowing Owls on Disturbed Sites." *Journal of Field Ornithology* 66:99–106.

45. Smith, B. W., and J. R. Belthoff. 2001. "Burrowing Owls and Development: Short-Distance Nest Burrow Relocation to Minimize Construction Impacts." *Journal of Raptor Research* 35:385–91.

46. Smith, M. D., and C. J. Conway. 2005. "Use of Artificial Burrows on Golf Courses for Burrowing Owl Conservation." *USGA Turfgrass and Environmental Research Online* 4:1–6.

47. Smith, M. D., C. J. Conway, and L. A. Ellis. 2005. "Burrowing Owl Nesting Productivity: A Comparison between Artificial and Natural Burrows on and off Golf Courses." *Wildlife Society Bulletin* 33:454–62.

48. Rebolo-Ifrán, N., M. Carrete, A. Sanz-Aguilar, S. Rodríguez-Martínez, S. Cabezas, T. A. Marchant, G. R. Bortolotti, and J. L. Tella. 2015. "Links between Fear of Humans, Stress and Survival Support a Non-Random Distribution of Birds among Urban and Rural Habitats." *Scientific Reports* 5:13723.

49. Bendire, C. E. 1892. "Life Histories of North American Birds." *U.S. National Museum Special Bulletin* 170:1–425.

50. Smith, M. D., and C. J. Conway. 2007. "Use of Mammal Manure by Nesting Burrowing Owls: A Test of Four Functional Hypotheses." *Animal Behaviour* 73:65–73.
51. Smith, M. D., and C. J. Conway. 2011. "Collection of Mammal Manure and Other Debris by Nesting Burrowing Owls." *Journal of Raptor Research* 45:220–28.
52. Levey, D. J., R. S. Duncan, and C. S. Levins. 2004. "Use of Dung as a Tool by Burrowing Owls." *Nature* 431:39.
53. Botelho, E. S., and P. C. Arrowood. 1996. "Nesting Success of Western Burrowing Owls in Natural and Human-Altered Environments." In *Raptors in Human Landscapes: Adaptations to Built and Cultivated Environments*, edited by D. Bird, D. Varland, and J. J. Negro, 61–68. San Diego: Academic Press.
54. Ray, J. D., N. E. McIntyre, M. C. Wallace, A. P. Teaschner, and M. G. Schoenhals. 2016. "Factors Influencing Burrowing Owl Abundance in Prairie Dog Colonies on the Southern High Plains of Texas." *Journal of Raptor Research* 50:185–93.
55. Todd, L. D. 2001. "Dispersal Patterns and Post-Fledging Mortality of Juvenile Burrowing Owls in Saskatchewan." *Journal of Raptor Research* 35:282–87.
56. Wellicome, T. I., and G. L. Holroyd. 2001. "The Second International Burrowing Owl Symposium: Background and Context." *Journal of Raptor Research* 35:269–73.
57. Environment Canada. 2012. "Recovery Strategy for the Burrowing Owl (*Athene cunicularia*) in Canada." *Species at Risk Act Recovery Strategy Series*. Ottawa, Ontario: Environment Canada.
58. Macías-Duarte, A., and C. J. Conway. 2015. "Distributional Changes in the Western Burrowing Owl (*Athene cunicularia hypugaea*) in North America from 1967 to 2008." *Journal of Raptor Research* 49:75–83.
59. Williford, D. L., M. C. Woodin, and M. K. Skoruppa. 2009. "Factors Influencing Selection of Road Culverts as Winter Roost Sites by Western Burrowing Owls." *Western North American Naturalist* 69:149–54.
60. Bartok, N., and C. J. Conway. 2010. "Factors Affecting the Presence of Nesting Burrowing Owls in an Agricultural Landscape." *Journal of Raptor Research* 44:286–93.
61. Scobie, C. A., E. Bayne, and T. Wellicome. 2014. "Influence of Anthropogenic Features and Traffic Disturbance on Burrowing Owl Diurnal Roosting Behavior." *Endangered Species Research* 24:73–83.
62. Arizona Game and Fish Department. 2008. "Burrowing Owl Project Clearance Guidance for Landowners." Unpublished Technical Report. Arizona Game and Fish Department, Phoenix, AZ.

CHAPTER 13

Peregrine Falcons: The Neighbors Upstairs

Joel E. Pagel, Clifford M. Anderson, Douglas A. Bell, Edward Deal,

Lloyd Kiff, F. Arthur McMorris, Patrick T. Redig, and Robert Sallinger

NOT ONLY AN ICONIC AERIAL PREDATOR with a long list of superlatives, the peregrine falcon (*Falco peregrinus*) continues to astound researchers and observers who delight in documenting its urban presence. However, not many of us who were interested in peregrine falcons during their population nadir would have predicted how adaptable they have become, with urban nest sites growing increasingly common.

To sustain their populations, most raptors require nest sites with protection from predation and persecution, and sufficient quantities of uncontaminated food. Populations of peregrine falcons declined drastically in many parts of their global range in the mid-20th century, primarily from the effects of dichlorodiphenyldichloroethylene (DDE), the principal metabolite of the pesticide dichlorodiphenyltrichloroethane (DDT).[1] High levels of DDE in blood and eggs are associated with eggshell thinning in peregrines and upper trophic level bird species. DDE causes shell thinning by inhibiting normal function of the shell gland, blocking calcium deposition on the outer layer of eggshells.[1,2,3,4,5] Abnormally thin eggshells can break during incubation, which lowers breeding success, eventually leading to population decline. In the United Kingdom and portions of Europe, the decline was attributed mainly to outright mortality caused by another pesticide, dieldrin/aldrin (HEOD), although DDE also caused problems for some birds there.[6] The relative importance of DDE, HEOD, and polychlorinated biphenyls (PCBs) to peregrine populations has been discussed in detail elsewhere.[7,8,9,10]

Peregrines populations have since rebounded, and peregrines now occupy nests not only in remote wilderness habitat but also in cities, arguably the most human-altered areas on the planet (color plate 11). The infrastructure humans build to support dense populations often arises near bodies of water, including rivers, lakes, or oceans. For a peregrine falcon, this creates suitable living space with ample diverse prey, structures with ledges for nests and perches, and open sky to hunt and soar.

Through antiquity, peregrines have been our neighbors, as their presence in urban areas likely began when humans first started building structures (buildings, bridges, towers, etc.) of sufficient size. Rothschild and Wollaston[11] reported peregrines on and around the Pyramids of Meroë in Sudan, and Fischer[12] noted peregrines on the Great Pyramids of Egypt. Other historic structures have hosted nesting peregrine falcons, including the Vatican (centuries of nesting records); Salisbury Cathedral, England; Heidelberg Castle, Germany; and the Sun Life building in Montreal, Quebec.[13,14,15,16] In 1996, Cade et al.[17] documented the numbers of urban peregrine falcon territories in North America ($n = 88$). Now with peregrine falcons occupying urban areas as small as Tarentum, Pennsylvania (pop. 4,500), and as large as New York City and other megapolises, a complete tally would be difficult. Contemporary observations such as a hungry subadult female peregrine perched on the top of the *California Screamin'* rollercoaster ride at Disneyland and an adult pair tandem-hunting pigeons on the Strip in Las Vegas suggest their common urban presence. Our cursory survey in 2017 revealed peregrines now nest on human-made structures in all but seven US states (Florida, Hawaii, Louisiana, Montana, South Carolina, Texas, and Wyoming). In Canada, structure-related nests occur in all provinces and territories, except Newfoundland, Nova Scotia, and Prince Edward Island. (We have no data for the Northwest Territories.) We were unable to collect similar data on urban presence from other countries and continents; however White et al.[16] mentioned urban nesting on structures by peregrines in cities in 26 countries on six continents, involving nine subspecies.

Resurgence and Human-Peregrine Urban History

We stand on the shoulders of our colleagues and friends who have amply documented urban peregrine falcons. The story of the five decades (1950–2000) of peregrine population decline and resurgence has been thoroughly described else-where.[15,16,17,18] Peer-reviewed articles, examining the city life and details of natural history by locale and subspecies, are too numerous to cite here; however, Herbert and Herbert,[19] Cade and Bird,[20] Cade et al.,[21] Drewitt,[22] and Caballero et al.[23] provide good introductions to urban peregrine falcons. In short, peregrines eat mostly birds (with exceptions), nest on flat surfaces on tall structures (with exceptions), and are capable of prodigious reproductive output (also with exceptions). From this base, the urban peregrine story takes several unexpected twists and turns.

At the population nadir when reintroduction techniques were being developed, there was concern in the United States regarding the rationale and priority of urban releases of peregrines. There were discussions on the merits and practicality of urban releases and how or even if they would benefit the recovery effort for remote (i.e., "wild") nesting peregrines. In time, urban centers were selected for reintroduction (hacking) and augmentation (fostering) because of ease of access, available staffing, paucity of predators, and opportunities for related outreach intended to promote this iconic species and its expensive recovery effort to the general public. Wildlife managers and raptor specialists in the midwestern and mid-Atlantic United States originally tried peregrine releases at natural sites but encountered excessive predation by great horned owls (*Bubo virginianus*) and raccoons (*Procyon lotor*), thus compelling the choice of urban release sites.

Over 7,000 captive-bred peregrines were released in the United States and Canada from the late 1970s through the early 1990s.[17] These urban peregrine releases created good public press wherever they happened. Positive newspaper and magazine accounts, documentaries, television interviews, and even game-show appearances featuring portable egg incubators or researchers wearing rubber-rimmed "copulation hats" used to collect peregrine semen helped get a conservation message to millions of people in multiple urban centers. Glowing responses from the public resulted from each reporter who experienced a fierce adult defending her nest or a cute, fuzzy nestling with seemingly oversized eyes (figure 13.1). Private citizens, politicians, and even wealthy benefactors, including Hollywood stars, sports celebrities, and CEOs were nudged or pulled into the peregrine world throughout North America and Europe via staged and spontaneous interactions with peregrine falcons. Urban peregrines gained followers, protectors, and notoriety.

Not all urban peregrines came from urban releases. In our respective study areas, we (specifically Pagel, Bell, McMorris, Redig, and Sallinger) noticed in the early 1990s, thanks to Visual Identification bands (colored aluminum alphanumeric bands made by ACRAFT; figure 13.1) placed on young in remote and

Figure 13.1. Public participation in banding urban peregrine falcon nestling. Colored VID bands are visible to left of rivet gun. Photo © Midwest Peregrine Society, courtesy of Jacquelyn Fallon.

wilderness nests, that many peregrines who established or occupied urban territories on the West Coast of North America originated from those nests far from urban areas. It soon became apparent in other North American locations that the remote population was aiding the growth of urban populations and vice versa. In the midwestern United States, urban releases and subsequent city nests were the source of the population that eventually reoccupied historical wild cliffs along the Mississippi River.

Human-Peregrine Urban Relationships

Peregrines surprise us with their nest-site selection and adaptability. We have observed nests on the tops of nuclear energy containment buildings and cooling towers, within bell towers of churches, next to sound cannons used to scare birds, within flowerpots on balconies of expensive high-rise apartments, on broken sacks of sand in a carillon tower high above a university campus, in the mast of a mothballed aircraft carrier, on operating shipbuilding cranes and cement counterweights on drawbridges (that move vertically whenever a large boat passes under the bridge; figure 13.2), and ground nests within protected urban California least tern (*Sternula antillarum browni*) colonies.[24] One site was even located in an old bald eagle *(Haliaeetus leucocephalus)* nest on the grounds of the Washington state governor's mansion. Bridge-nesting peregrines are especially interesting (figure 13.3). We have discovered nest scrapes on narrow open I-beams, deep (20 m) inside hollow concrete bridge supports, and at junctures of steel I-beams. In some cases, the eggs and young are constantly vibrated by trucks and cars passing right over, or under the site or the scrapes are set so far back in a bridge support that the young are raised in 24-hour darkness. We

Figure 13.2. Enclosed nest box used to move peregrine falcons from their original nest on a moving vertical lift span counterweight in Portland, Oregon. Photo © Audubon Society of Portland, courtesy of Robert Sallinger.

Figure 13.3. Adult peregrine falcon perching on newly constructed San Francisco–Oakland Bay Bridge, California. Photo © Mary Malec.

have also noted nests on the top of bare cement bridge columns with little to no substrate to cushion the eggs (figure 13.4) or laid atop a thick substrate of sharp vole bones from barn owl (*Tyto alba*) pellets. Remarkably, most of these have resulted in nesting success.

Urban releases, the influx of peregrines from remote areas, and their subsequent nests in cities also created enemies. Some pigeon (*Columba livia domestica*) enthusiasts still rue the presence of peregrines in urban ecosystems, and among their ranks were vigilantes who killed hundreds of raptors, including peregrine falcons, which were classified as federally endangered in the United States at that time. In the early 1990s, bullet holes were found near nest scrapes on several skyscrapers in Los Angeles, ostensibly from sharpshooters aiming for peregrines. In 2007, a 14-month undercover federal investigation called *Operation High Roller* resulted in the arrests of 16 defendants in California, Oregon, Texas, and Washington who had carried out systematic persecution (shooting, poisoning, beating, mutilation, etc.) of raptors, including peregrine falcons. Federal investigators estimated leaders and members of pigeon enthusiast organizations in Los Angeles, California, were responsible for killing 1,000 to 2,000 raptors annually, with additional estimates for Portland, Oregon, of more than 1,500 raptors killed, including urban peregrine falcons.

Building managers and residents sometimes complain about excreta ("whitewash") stains on building facades and windows or adult peregrines rendering prey near conference room or apartment windows. Because of specialized

Figure 13.4. Peregrine falcon nest scrape on bare cement bridge column in Portland, Oregon. Photo © Joel E. Pagel.

skills and permits, Pagel was asked to climb to peregrine plucking posts on apartment buildings in front of windowed ledges to remove piles of prey remains that were left by satiated peregrines and to calm angry building tenants who were frustrated by peregrines killing "their building's" pigeons or defecating on "their" window ledges. Peregrine researchers have even been asked to help solve tense situations in which building managers believed satanic rituals by humans had occurred near the base of their buildings due to large amounts of decapitated and plucked bodies of pigeons and other birds. Many of these situations can be humorous to those of us who know peregrines but are very serious to tenants, structure managers, and security officers.

Conversely, urban peregrines have given numerous laypeople the opportunity to contribute to research efforts to better understand falcon ecology. Several important factors—such as "easy" nest access for banding, the use of Visual Identification bands, and the advent of digital photography—have created an improved understanding of individual peregrines. Many peregrine nests have been monitored through the windows of sumptuous boardrooms and lavish corner offices hundreds of meters above the city streets; likewise, numerous buildings have "adopted" their birds by having live video streams of "their" nests available in the lobby and on the internet. As a result, we now have improved data on peregrine longevity, peculiarities, conflicts, movements, and reproductive output not ordinarily available at remote nests. We have also noted inbreeding at urban nests where siblings from the same clutch eventually refound each other and successfully bred at another urban site.[25] We (authors Pagel, Anderson, Deal, and Sallinger) were fortunate to follow the entire life of an individual peregrine from hatching near Portland, Oregon, to becoming a breeding adult in Seattle, Washington, 235 km north. She was followed through 14 years of reproductive life, with almost all of her young banded, revealing her genetic contribution to that city's population. Similar data and observations made on peregrines in the midwestern United States are available on a publicly accessible database (http://www.midwest peregrine.org), showcasing over 30 years of individual birds and nest lineages.

Because of urban-nesting falcons, researchers have more knowledge of their behavior and ecology. For example, we now know peregrine falcons in North America and Europe hunt near their eyries at night, aided by city lighting,[26,27,28] and in northern-latitude cities known to be very cold, peregrines overwinter instead of migrating.[29] In the Pacific Northwest climates of Portland, Oregon, and Seattle, Washington, resident peregrines have been observed copulating, presumably maintaining their pair bond, in every month of the year. In Portland, Oregon, urban peregrines are annually anticipated at autumn concentrations of migrant Vaux's swifts (*Chaetura vauxi*) that roost by the thousands in an abandoned chimney at the Chapman Elementary School, with hundreds of humans picnicking below and cheering for the swifts or for the peregrines.

Current Issues with Urban Peregrines

Despite urban living, peregrine falcons are not spared from the myriad issues affecting individual survival. Some hazards are chronic and require ongoing human intervention: bridge peregrine falcon young falling into the water during fledging attempts;[30] radiant heat from bridge girders baking eggs;[31] grounded nestlings and fledglings run over by traffic (bicycles, cars, trucks, and trains); parasitic *Trichomonas* infestations transferred from rock pigeons; and secondary poisoning during pigeon, corvid, and European starling (*Sturnus vulgaris*) culls. Urban peregrine falcons are still susceptible to disturbance (e.g., Hollywood movie crews, suicidal bridge jumpers, window washers, presidential motorcades, bomb squad drills), window and glass wall strikes, electrocution and collisions with utility distribution lines, and falling down into the shafts of air conditioners and heating ducts. And as also happens at remote nests, territorial skirmishes or usurpation can turn deadly for urban adult peregrines.

Conflicts surrounding regularly scheduled maintenance, construction, or even human safety (diving peregrines sometimes strike unwary workers or tenants; figure 13.5) are additional issues. All of the authors have met corporate lawyers and building managers concerned with liability from peregrine prey being dropped on unsuspecting pedestrians, small bits of nest-box gravel pockmarking parked cars or breaking inclined window panes below the nests, or attacks on window washers and building occupants. We have found that starting early before peregrine courtship begins and working cooperatively with maintenance staff affords positive outcomes to even the most difficult situations. Observation data are critical; knowing ranges of dates of courtship, egg laying, hatching, and fledging can provide the structure managers with sufficient information to pre-plan work activity. Using these approaches, we have rarely had to remove nest boxes or block nesting, even with proven dangers to humans or peregrines, and have often shown that routine maintenance can generally be scheduled around a peregrine nesting chronology. We have found that explaining peregrine biology and nesting behavior, providing recommendations of when maintenance or construction can be undertaken safely, and pointing out the benefits of positive press will convince almost all building and bridge managers to stay within the boundaries of the law to protect nesting peregrines.

Figure 13.5. Adult peregrine falcon defending her nestlings during banding at the Rachel Carson State Office Building, Harrisburg, Pennsylvania. Photo © Joe Kosack.

We have also encountered building and bridge managers who request recommendations on how to *attract* peregrine falcons to their buildings and bridges or make current nesting situations better to control starlings and pigeons. Although we have no quantitative data, some building and bridge managers have indicated that the presence of peregrine falcons on their structures has lessened maintenance costs via reducing the burden of corrosive bird excrement. To attract peregrines to structures and make poor nests better, we have designed and built nest trays and carried heavy loads of pea gravel to favored ledges to provide more permanent nesting substrate (figures 13.2, 13.6, 13.7).

The incidence of rescue and rehabilitation of peregrines has increased in proportion to the growth of urban populations (figure 13.8). In numerous instances, observers and the public have bonded with injured or accident-prone peregrines, increasing the empathetic attitude toward their urban birds. In Oregon and Washington, nestlings from precarious bridges were taken at older ages, or after falling from bridges, and transferred to falconers who were allotted a lottery permit for state-sanctioned falconry. This has reduced resource conflict and potential for human-caused disturbance at remote nest sites.

Figure 13.6. Adult peregrine falcon on bridge prior to the banding of nestlings near her pea gravel-enhanced nest. Photo © Audubon Society of Portland, courtesy of Robert Sallinger.

Thoughts on the Future

As conservationists, we have used the stature and popularity of peregrine falcons to enhance public understanding of endangered species, natural ecosystems, and the universal need for connecting to wild things. From conservation nonprofits that team up with local businesses to host peregrine falcon festivals to public involvement through Peregrine Patrols or Fledge Watch Squads, the iconography of the species has a seemingly universal appeal. Peregrines provided urbanites with tangible opportunities to assist in ongoing protection without traveling to distant lands or remote wildernesses. Often, we are pulled aside by parents who relate how their child desperately wants to study raptors and that peregrine falcons are his or her favorite animal. In patient tones to bright-eyed children and adults, we explain tomial teeth, spiral dives, eggshell

Figure 13.7. Peregrine falcon nest tray on building in Seattle, Washington. Photo ©
Edward Deal.

thinning, and predator-prey relationships (figure 13.1). We use the success of
peregrine conservation to teach about other raptors, predators, and all other
species, using their multicontinental resurgence as the intellectual thread to
connect to other species and spaces, and as an inspiration for work on correla-
tive conservation problems.

Peregrines will continue to amaze us as they nest in more cities and towns
throughout the world (figures 13.9, 13.10). We acknowledge that in some
instances they may create issues involving predation of special-status species or impact structure repair and maintenance schedules and that, for some
human residents, peregrines may cause perceptional discomfort via the maca-
bre or distasteful observation of the raptor killing and eating a backyard bird
or park pigeon. Yet having seen the species at the brink of extinction, we relish
the opportunity of working with urban peregrine falcons to help other humans
experience the magnificence and grandeur of the fastest and, not so very long
ago, one of the most endangered birds on our planet.

Figure 13.8. Second attempt at flight; hacking previously grounded peregrine falcon fledglings from the top of their natal building at Fashion Island, Irvine, California. Photo © D. Gollwitzer.

Figure 13.9. Flyby of adult female peregrine falcon to protect nestlings in city center, Minneapolis, Minnesota. Photo © Midwest Peregrine Society, courtesy of Jacquelyn Fallon.

Figure 13.10. Urban nest attendance of peregrine falcon nestling. Photo © Midwest Peregrine Society, courtesy of Jacquelyn Fallon.

Acknowledgments

This account has benefited by the review, comments, and encouragement of numerous people including K. T. Cleveland, M. Durflinger, B. A. Millsap, J. Sipple, and guidance and review by C. Boal and C. Dykstra. K. Gunther assisted with the tallying of urban peregrine presence in the United States and Canada. We thank C. Ellingson, J. Fallon, D. Gowlitzer, J. Kosack, and M. Malec for the use of their photographs. We also thank the hundreds of peregrine observers who provided observations, picked up grounded nestlings, or kept us company during field efforts and the numerous structure managers who have facilitated successful nest sites in urban areas under unusual conditions. The findings and conclusions in this article are those of the authors and do not necessarily represent the views of the US Fish and Wildlife Service.

Literature Cited

1. Peakall, D. B., and L. Kiff. 1979. "Eggshell Thinning and DDE Residue Levels among Peregrine Falcons (*Falco peregrinus*): A Global Perspective." *Ibis* 121:200–204.

2. Bitman, J., H. C. Cecil, and G. G. Fries. 1970. "DDT-Induced Inhibition of Avian Shell Gland Carbonic Anhydrase: A Mechanism for Thin Eggshells." *Science* 168:592–94.

3. Peakall, D. B. 1970. "p,p'-DDE: Effect on Calcium Metabolism and Concentration of Estradiol in the Blood." *Science* 1698:592–94.

4. Peakall, D. B. 1975. "Physiological Effects of Chlorinated Hydrocarbons on Avian Species." In *Environmental Dynamics of Pesticides*, edited by R. Hague and V. Freeds, 343–60. New York: Plenum Press.

5. Miller, D. S., W. B. Kinter, and D. B. Peakall. 1976. "Enzymatic Basis for DDE-Induced Eggshell Thinning in a Sensitive Bird." *Nature* 259:122–24.

6. Nisbet, I. C. T. 1988. "The Relative Importance of DDE and Dieldrin in the Decline of Peregrine Falcon Populations." In *Peregrine Falcon Populations: Their Management and Recovery*, edited by T. J. Cade, J. H. Enderson, C. G. Thelander, and C. M. White, 351–75. Boise, ID: The Peregrine Fund.

7. Cade, T. J., J. H. Enderson, C. G. Thelander, and C. M. White, eds. 1988. *Peregrine Falcon Populations: Their Management and Recovery*. Boise, ID: The Peregrine Fund.

8. Risebrough, R. W., and D. B. Peakall. 1988. "Commentary—the Relative Importance of Several Organochlorines in the Decline of Peregrine Falcon Populations." In *Peregrine Falcon Populations: Their Management and Recovery*, edited by T. J. Cade, J. H. Enderson, C. G. Thelander, and C. M. White, 449–62. Boise, ID: The Peregrine Fund.

9. Peakall, D. B., and J. L. Lincer. 1996. "Do PCBs Cause Eggshell Thinning?" *Environmental Pollution* 91:127–29.

10. Newton, I. 2013. "Organochlorine Pesticides and Birds." *British Birds* 106:189–205.

11. Rothschild, N. C., and A. F. R. Wollaston. 1902. "On a Collection of Birds from Shendi, Sudan." *Ibis* 44:1–33.

12. Fischer, W. 1967. *Der Wanderfalk*. Wittenberg, Germany: A. Ziemsen Verlag.

13. Olivier, G. 1953. "Nidificiation du Faucon Pelerine sur les Edifices." *L'Oiseau et la Revue Française d'Ornithologie* 23:109–204.

14. Hickey, J. J., and D. W. Anderson. 1969. "The Peregrine Falcon: Life History and Population Literature." In *Peregrine Falcon Populations: Their Biology and Decline*, edited by J. J. Hickey, 3–42. Madison: University of Wisconsin Press.

15. Ratcliffe, D. A. 1993. *The Peregrine Falcon*. 2nd ed. London, UK: T. and A. D. Poyser.

16. White, C. M., T. J. Cade, and J. H. Enderson. 2013. *Peregrine Falcons of the World*. Barcelona, Spain: Lynx Edicions.

17. White, C. M., N. J. Clum, T. J. Cade, and W. G. Hunt. 2002. "Peregrine Falcon (*Falco peregrinus*)." In *The Birds of North America*, edited by P. G. Rodewald. Ithaca: Cornell Lab of Ornithology. Accessed April 10, 2017. https://birdsna.org/Species-Account/bna/species/perfal.

18. Cade, T. J., and W. Burnham, eds. 2003. *Return of the Peregrine*. Boise, ID: The Peregrine Fund.

19. Herbert, R. A., and K. G. S. Herbert. 1965. "Behavior of Peregrine Falcons in the New York City Region." *Auk* 82:62–94.

20. Cade, T. J., and D. M. Bird. 1990. "Peregrine Falcons, *Falco peregrinus*, Nesting in an Urban Environment: A Review." *Canadian Field-Naturalist* 104:209–18.

21. Cade, T. J., M. Martell, P. Redig, G. Septon, and H. Tordoff. 1996. "Peregrine Falcons in Urban North America." In *Raptors in Human Landscapes: Adaptations to Built and Cultivated Environments*, edited by D. Bird, D. Varland, and J. J. Negro, 3–13. San Diego: Academic Press.

22. Drewitt, E. 2014. *Urban Peregrines*. Exeter, UK: Pelagic Publishing.

23. Caballero, I. C., J. M. Bates, M. Hennen, and M. V. Ashley. 2016. "Sex in the City: Breeding Behavior of Urban Peregrine Falcons in the Midwestern US." *PLoS ONE* 11(7): e0159054. doi:10.1371/journal.pone.0159054.

24. Pagel, J. E., R. T. Patton, and B. Latta. 2010. "Ground Nesting of Peregrine Falcons (*Falco peregrinus*) near San Diego, California." *Journal of Raptor Research* 44:323–25.

25. Pagel, J. E., and J. Sipple. 2011. "Incident of Full Sibling Mating in Peregrine Falcons (*Falco peregrinus*)." *Journal of Raptor Research* 45:97–98.

26. Wendt, A., G. Septon, and J. Moline. 1991. "Juvenile Urban-Hacked Peregrine Falcons (*Falco peregrinus*) Hunt at Night." *Journal of Raptor Research* 25:94–95.

27. Rejt, L. 2004. "Nocturnal Feeding of Young by Urban Peregrine Falcons (*Falco peregrinus*) in Warsaw (Poland)." *Polish Journal of Ecology* 52:63–68.

28. DeCandido, R., and D. Allen. 2006. "Nocturnal Hunting by Peregrine Falcons at the Empire State Building, New York City." *Wilson Journal of Ornithology* 118:53–58.

29. Septon, G. 2000. "Overwintering by Urban-Nesting Peregrine Falcons (*Falco peregrinus*) in Midwestern North America." In *Raptors at Risk: Proceedings of the V World Conference on Birds of Prey and Owls*, edited by R. D. Chancellor and B.-U. Meyburg, 455–61. Berlin, Germany: World Working Group on Birds of Prey and Owls; Surrey, BC: Hancock House Publishers.

30. Bell, D. A., D. P. Gregoire, and B. J. Walton. 1996. "Bridge Use by Peregrine Falcons in the San Francisco Bay Area." In *Raptors in Human Landscapes: Adaptations to Built and Cultivated Environments*, edited by D. Bird, D. Varland, and J. J. Negro, 15–24. San Diego: Academic Press.

31. Hurley, V. 2013. "Factors Affecting Breeding Success in Peregrine Falcons (*Falco peregrinus macropus*) across Victoria 1991–2012." PhD thesis, School of Life and Environmental Sciences, Deakin University.

PART III

Conservation and Management

CHAPTER 14

Raptor Mortality in Urban Landscapes

James F. Dwyer, Sofi Hindmarch, and Gail E. Kratz

D URING THE NORMAL COURSE OF their lives in exurban landscapes, raptors are killed by weather, accidents, disease, predators, and each other. For example, when late winter storms bring low temperatures, snow, and ice, spring migrants or their nestlings can rapidly deplete limited energetic reserves. Summer storms and monsoons can bring high winds, torrential rain, and hail, adding risks of falling rock and ice (which can dislodge when melting) crushing nestlings, breaking branches, and collapsing nests. Other raptors die of dehydration and hyperthermia in arid exurban environments, particularly nestlings in south-facing nests. Accidents in flight occur when young raptors are learning to fly and when adult raptors are focused on prey or competitors and thus fail to perceive an upcoming obstacle. Disease can result from exposure to a wide variety of vectors, including infected prey, mosquitoes, and social transmission. Predators most often consume nestlings in nests, but larger raptors also prey on smaller species, and intraspecific and interspecific competition routinely lead to injury or death. Across these factors, starvation, which is typically an indirect effect, is often the proximate cause of death in exurban environments.

Urban areas include all of these dangers in addition to mortality risks related to human-influenced landscapes. For example, more spatially cluttered environments can increase collision risk, concentrated prey and water sources can increase exposure to disease vectors, and power lines introduce electrocution risk. Despite these dangers, expanding urbanization, reduced human persecution, habituation to human activity, and use of anthropogenic resources is increasingly leading raptors to disperse to, and nest in, urban areas.[1,2] Urban nesting by raptors has become so common that it is now noteworthy primarily when a species transitions to the behavior, as northern goshawks (*Accipiter gentilis*) did in Europe in the 1970–80s,[3] merlins (*Falco columbarius*) did in the central prairies of North America in the 1980s,[4,5] and as crested caracaras (*Caracara cheriway*) may be doing now.[6]

Nesting in urban areas can provide raptors with predictable high quality food, both for specialists like Cooper's hawks (*Accipiter cooperii*)[2,7] and peregrine falcons (*Falco peregrinus*)[8] and for generalists like burrowing owls (*Athene cunicularia*) and barred owls (*Strix varia*).[9] Urban nesting can also provide stable nesting territories[10,11] and reduce predation.[12,13] For raptors habituated to urban landscapes, these benefits can facilitate high reproductive outputs compared to natural habitats.[13] Though productivity can be high, it is not necessarily so, and survival can be reduced across age classes by a suite of anthropogenic causes of mortality.[14,15,16]

In this chapter, we discuss causes of mortality of urban raptors with a focus on anthropogenic factors, because these can be particularly localized and prevalent. Nevertheless, many of the causes of mortality described here can also affect exurban raptors. Thus this chapter, which is organized from most to least important for urban raptor populations, serves not only as a summary of urban mortality, but also as a partial summary of causes of raptor mortality in general. Importantly, high concentrations and unique combinations of urban causes of mortality may act synergistically to create novel risks, regardless of the quality of the individual raptor involved. For example, in many species, individuals of all combinations of age and sex classes routinely perch on overhead electric power structures.[15] Occasionally, an individual is electrocuted due to perch location on a particular pole without regard to aspects of its fitness.[17]

Collision

Collisions appear to be the single greatest cause of urban raptor mortality.[18] Collisions with vehicles have the greatest breadth, though this may in part reflect detection bias if carcasses near roads are more likely to be noticed.[19] In one study, collisions involved 73 percent of urban Accipitriforme and Falconiforme species and 63 percent of urban Strigiforme species considered, and accounted for 39 percent and 32 percent of mortalities, respectively.[18] As highlighted in chapter 2, raptors involved in collisions with vehicles may have been hunting in the open areas adjacent to roads, scavenging the carcasses of other animals previously struck by vehicles, or simply transecting a road during normal movement.[20,21] Designing roads with a barrier such as fence or hedge that redirects raptors to fly higher when crossing may reduce mortality for some species.[22] Other mitigation options involve reducing the speed and volume of traffic and increasing the awareness of motorists, but efforts to implement these changes have often been met with resistance.[23]

To a lesser extent, urban collisions with moving objects also include aircraft. Raptor-specific mitigation strategies to avoid these collisions focus on removing prey and nesting opportunities from within airport grounds.[24] Collisions with wind turbines also occur, but because few large-scale wind farms are in urban areas, most mortality of urban raptors does not include this factor, and effective mitigation strategies specific to urban raptors have not been identified.

Collisions with stationary objects most often include windows (figure 14.1A) when a raptor pursues prey through cluttered environments; accipiters and strigiformes are particularly prone to these collisions.[18] These collisions can be especially prevalent where passerine bird feeders are in close proximity to windows to facilitate observation from inside buildings.[25] Mitigation generally involves shifting feeders away from windows and adding stickers to windows to increase their visibility. Because of the limited success of these approaches, ongoing research is investigating alternate mitigation strategies such as using reflected ultraviolet light—which birds see and people do not—from stickers, etchings, and glass.[26]

Other less-prevalent stationary objects in urban areas with which raptors collide include power lines, wire fences (figure 14.1B), and guy wires to radio and cell towers. Collision mitigation for wires typically involves installation of devices designed to increase the visual prominence of the wire, with recent advances particularly focused on mitigating nocturnal collision.[27]

The most common indicator of collision mortality is the circumstantial location of a raptor directly below a window, wire, or other object. Occasionally, this evidence can be bolstered with observations of a bird-shaped smear on the window or feathers stuck to the object involved.

Electrocution

Electrocution appears to be the second-most prevalent agent of anthropogenic mortality for urban raptors, accounting for an estimated 48 percent of urban

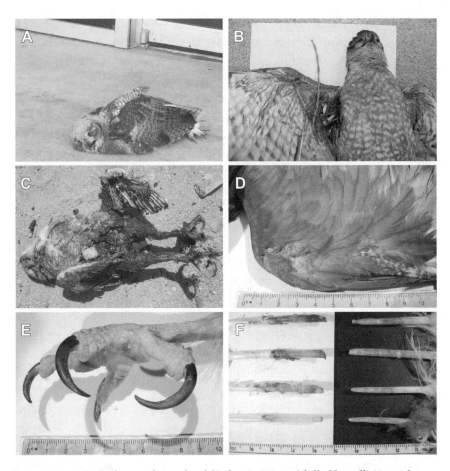

Figure 14.1. *A*, Juvenile great horned owl (*Bubo virginianus*) killed by collision with a residential window; *B*, Adult Cooper's hawk (*Accipiter cooperii*) killed when entangled on an urban barbed wire fence; *C*, Extensive charring of an electrocuted adult great horned owl; *D*, Minor charring on the wing of an adult Harris's hawk (*Parabuteo unicinctus*); *E*, Petechial hemorrhage (red streak) along the foot and leg of an electrocuted adult Harris's hawk; *F*, Flight feathers molted as a result of electric shock injury (*left*) and molted normally (*right*) in an adult red-tailed hawk (*Buteo jamaicensis*).

Accipitriforme and Falconiforme and 5 percent of urban Strigiforme deaths.[18] Avian electrocutions have been documented most often where overhead electric systems occur in nonforested landscapes. In these cases, utility poles are often the tallest perching substrate available. Because most of the energized components on overhead electric systems are not insulated, electrocutions result when a raptor simultaneously contacts two differently energized components.[28] Larger

species and larger individuals, particularly females, are disproportionately electrocuted because the larger the bird, the greater the likelihood of simultaneous
contact.[15,29] In urban habitats, electrocutions occur most often on complex poles
near nests.[15]

Symptoms range from obvious burns, which can cover the entire body (figures 14.1C through 14.1F), to injured or dead birds with minimal or no obvious external injuries.[30] Some injured individuals recover in the wild,[29] but more
often injured birds in urban areas are reported to wildlife rehabilitation facilities,
which attempt to treat them (see chapter 16), often with limited success.[31]

Poisoning

Exposure to chemical contaminants can be high in urban environments due to
construction, industrial activities, and vehicle exhaust.[32,33,34] Because raptors tend
to be at the top of their food chains, bioaccumulation can make them particularly susceptible to contaminants. Historically, dichlorodiphenyltrichloroethane
(DDT), a persistent organic pollutant, was the most important contaminant to
raptor populations in North America and Europe. Since DDT was banned in
the 1970s and 1980s in most developed countries, many populations of the most
affected species (e.g., bald eagles [*Haliaeetus leucocephalus*], ospreys [*Pandion
haliaetus*], and peregrine falcons) have recovered.

Dichlorodiphenyldichloroethylene (DDE), a persistent, bioaccumulative
breakdown product of DDT, and polychlorinated biphenyls (PCBs) are still present in sediment and biota, and use of other contaminants such as flame retardants
and rodenticides is ongoing.[35] This section provides a brief introduction to each of
these contaminants, but because poisoning symptoms and treatments are highly
variable depending on the dose and compound involved, symptoms and treatments are not described here.

The most common PCBs are polychlorinated dibenzo-p-dioxins (PCDDs)
and polychlorinated dibenzofurans (PCDFs). These may cause low reproductive output in raptors,[36] but linking high PCB concentrations in eggs to reduced
fecundity is challenging due to nearly ubiquitous cooccurrence with DDE.[37] Sublethal reproductive impacts have been documented in bald eagles and ospreys.[38,39]

Flame-retardant polybrominated diphenyl ethers (PBDEs) are used in plas-
tics, electronics, and textiles. PBDE residues have been detected in the eggs of
raptors residing in or close to urban environments, such as ospreys,[40] Cooper's
hawks,[35] peregrine falcons,[41] and little owls (*Athene noctua*).[42] The highest levels
of PBDEs in raptors have been documented in urban individuals using landfills,

where PBDEs are released as treated products decompose. PBDEs can mimic thyroid hormones and affect neural systems in laboratory animals,[43] but despite high PBDE concentrations detected in urban raptors, the degree to which PBDEs impact productivity or have other sublethal effects remains unknown.[35]

Anticoagulant rodenticides (ARs), particularly second-generation anticoagulant rodenticides (SGARs), are used extensively in urban environments to suppress rodent populations. By design, ARs and SGARs are highly toxic compounds intended to kill after a single ingested dose, are slow-acting to avoid target species associating baits with risk, and have a long biological half-life in the liver (e.g., Brodifacoum mouse liver $t_{1/2}$ = 307 d).[44] Nonmigratory raptors feeding on rodents, particularly AR-targeted rodents, in urban environments are disproportionally affected by ARs and SGARs. In some species, ARs have been detected in 75–100 percent of the raptor populations examined.[45,46] Raptors appear to survive some AR and SGAR exposure, but correlations between measures of internal toxin levels, sublethal effects, and population effects are unknown.[47]

Several aspects of detecting and diagnosing contaminant exposure and poisoning create challenges for understanding effects. For example, evidence of poisoning in carcasses may be more difficult to detect than other causes of mortality, requiring specific diagnostic testing that is expensive and not always available. Exposure to contaminants also may cause an individual to be more susceptible to other risks, requiring researchers to distinguish proximate from ultimate causes of mortality. Poisoning diagnoses can also be complicated by the "cocktail effect" of exposure to multiple compounds, which is often the case for urban raptors. Lead and mercury from industry, vehicle exhaust, and landfills are also important chemical contaminants with the potential to affect raptors.[36,48,49]

Disease

Infectious diseases are important agents of mortality in urban raptors because urban environments include microhabitats that can serve as reservoirs for pathogens.[19] For example, anthropogenic standing water in birdbaths, ponds, and ditches in parks, greenbelts, and around residences can support mosquito (*Culicidae* species) populations that act as transmitting agents (vectors) of various pathogens. Standing water also can facilitate disease transmission because water bodies may attract both raptors and their prey, particularly within urban contexts where backyard feeders and structurally complex landscaping create ample foraging opportunities for both. As described in chapter 4, standing water may be entirely a result of urbanization in areas where standing water was not present naturally. This likely introduces types or levels of disease risk not naturally present. In North America, increasingly common backyard poultry in urban areas may exacerbate risks of disease transmission because personal (hobby) flocks are associated with standing water and food but are rarely protected by biosecurity protocols designed to mitigate exposure to disease. Consequently, personal flocks can create the potential for development and persistence of urban reservoirs of avian disease. Coupled with reductions in the severity and duration of winter conditions associated with global climate change, urban environments also have begun to provide year-round habitat for wild birds that historically migrated. Urban environments may also support atypically dense raptor populations.

Several avian or zoonotic diseases are common or developing in urban habitats. Three of these—avian influenza, avian pox, and trichomoniasis—have been particularly deleterious for infected raptors and are described here as examples. More detailed descriptions of these and other diseases in raptors, including specific clinical signs, are well described in texts for veterinary and wildlife rehabilitation care.[50,51]

Avian influenza (family *Orthomyxoviridae*) describes infection with avian influenza type A viruses. Raptors contract avian influenza primarily through consumption of infected prey.[52] Pathogenicity varies, but highly pathogenic avian influenza (HPAI) strains can cause high mortality among infected raptors.

Avian pox describes infection with an avipox virus (family *Poxviridae*). These highly contagious infections are transmitted through mosquito vectors and by virus particles shed from pustules in the skin and mouth.[53] Avian pox infections can cause diphtheria-like symptoms including pustules in the skin, mucous membranes, and the upper respiratory tract—the last being most associated with mortality. Seroprevalence analyses indicate many raptors, particularly those that primarily hunt other birds, have been exposed to avian pox and survived.[53]

Trichomoniasis is caused by a flagellated trichomonas protozoan (*Trichomonas gallinae*) ingested with contaminated prey or water.[54] Trichomoniasis is particularly prevalent in the nestlings of urban-nesting accipiters, as highlighted in chapter 4, including Cooper's hawks and Eurasian sparrowhawks (*Accipiter nisus*).[55,56] Trichomonas enters through the mouth and nose and can colonize the eye and brain. Advanced stages in raptors are characterized by growths in the mouth (figure 14.2A), sinuses, and upper gastrointestinal tract that cause difficulty swallowing and breathing. Cause of death is typically a combination of starvation, dehydration, and suffocation.[54]

Prevention for the diseases described here mostly involves widespread and consistent mosquito control and more careful consideration of potential

Figure 14.2. *A*, Juvenile Cooper's hawk (*Accipiter cooperii*) with trichomonas (*Trichomonas gallinae*) lesions on the tongue and the roof of the mouth; *B*, Osprey (*Pandion haliaetus*) nest incorporating substantial anthropogenic materials; *C*, Juvenile osprey entangled in twine in the nest; *D*, Gunshot wound in the lower ventral abdomen of an adult golden eagle (*Aquila chrysaetos*) killed when perched near an urban area.

microhabitats created by urban bird feeders and landscaping. Each of these are often prohibitively expensive or resisted by human populations, and consequently, the avian diseases described here and many others, like West Nile virus (*Flaviviridae* species), are likely to persist in urban habitats.[50,51]

Other Impacts: Predation, Disturbance, Trash, and Persecution

Predation appears to be generally reduced in urban areas.[12,13] However, in some cases, nest predation by unusually dense populations of urban mammals may offset reductions in predation by non-urban species, reducing or reversing the reduction in predation pressure overall.

Disturbance can lead to nest abandonment even in human-tolerant species.[57] If nests are active (i.e., eggs were laid) when abandonment occurs, then egg or nestling mortality follows. Mortality of nestlings also can occur when tree trimming destroys active nests or when nests are constructed on inappropriate substrates. For example, when bald eagles nest in species of *Eucalyptus* trees, the weight of the nest can collapse the tree (E. Mojica, pers. comm.). When peregrine falcons nest on the iron superstructures of bridges, solar heat radiating from the iron can overheat the eggs.[58] Nestling mortality also occurs when urban raptors incorporate nest materials that can be dangerous to nestlings. For example, urban-nesting ospreys often bring discarded fishing line or twine to nests (figures 14.2B and 14.2C), sometimes resulting in entanglement of their nestlings.[59]

As will be explored in chapter 15 of this volume, raptors thought to be threats to human residents, domestic animals, or passerines at feeders also may be shot (figure 14.2D). Generally, the prevalence of shooting appears to be declining in urban areas, though exurban persecution persists in some regions.[60,61,62]

Mitigation for these factors is generally not implemented. Tree trimming can be avoided during breeding seasons, but doing so requires knowledge of, and interest in, the welfare of urban-nesting raptors, criteria that are often unmet without the involvement of local biologists. Avoiding unsafe disposal of fishing line or twine is an obvious solution to entanglement, but consistent implementation remains to be achieved.

Conclusions

This chapter provides a brief summary of some agents of mortality affecting urban raptors. Sometimes, the death of an urban raptor is straightforward with a single cause of mortality. However, in many cases, novel levels and combinations of risks that are present in urban settings make distinguishing proximate from ultimate causes of mortality challenging. Identification of biologically meaningful effects can also be difficult. For example, as explored in chapter 9 of this volume, Harris's hawks (*Parabuteo unicinctus*) nest in urban areas in southern Arizona. Although urban electrocution mortality has been hypothesized to affect social ecology in Harris's hawks and act as an ecological trap in Tucson, Arizona,[17,63] the species appears to be thriving, increasing from about 10 breeding pairs in 1975[64] to 46 pairs in 1993[63] and 62 pairs in 2003.[15]

The relative magnitude of various causes of mortality in urban raptors differs across landscapes and species. In forested urban areas, homes set into woodlands may have higher levels of window collisions. In urban areas interspersed with agriculture, mortality may be primarily due to pesticide poisoning. Due to this habitat- and behavior-mediated variability, conservation and management actions require site-specific approaches. As will be addressed in chapter 16, personnel interested in understanding specific incidents should seek advice from local raptor rehabilitators who will be familiar with the symptoms, signs, and patterns of locally common causes of mortality.

Throughout this chapter, the anthropogenic-induced mortalities described are usually accidental but are nevertheless illegal in many areas. For example, in the United States, the Migratory Bird Treaty Act stipulates strict liability for killing or wounding any migratory bird, regardless of whether a person knowingly or intentionally commits the act. Consequently, most of the causes of mortality described here are violations of law, though prosecutions have been limited to deliberate shooting and to electrocutions on structures where risk was known, but not mitigated, at least in the United States.

Acknowledgments

We thank Judy Scherpelz and Michael Tincher of the Rocky Mountain Raptor Program and Michelle Willette of the Raptor Center at the University of Minnesota for insights into the mortality of urban raptors. Photos courtesy of Michael C. Tincher (figure 14.1*A*; figure 14.2*A*, *C*, *D*), Dianna Flynt (figure 14.1*B*), and James F. Dwyer (all others).

Literature Cited

1. Gehlbach, F. R. 2012. "Eastern Screech-Owl Responses to Suburban Sprawl, Warmer Climate, and Additional Avian Food in Central Texas." *Wilson Journal of Ornithology* 124:630–33.

2. Boggie, M. A., and R. W. Mannan. 2014. "Examining Seasonal Patterns of Space Use to Gauge How an Accipiter Responds to Urbanization." *Landscape and Urban Planning* 124:34–42.

3. Rutz, C. 2008. "The Establishment of an Urban Bird Population." *Journal of Animal Ecology* 77:1008–19.

4. James, P. C. 1988. "Urban Merlins in Canada." *British Birds* 81:274–77.

5. Warkentin, I. G., N. S. Sodhi, R. H. M. Espie, A. F. Poole, L. W. Oliphant, and P. C. James. 2005. "Merlin (*Falco columbarius*)." In *The Birds of North America*, edited by P. G. Rodewald. Ithaca: Cornell Lab of Ornithology. Accessed July 27, 2017. https://birdsna.org/Species-Account/bna/species/merlin.

6. Dwyer, J. F., and J. P. Dalla Rosa. 2015. "Use of Anthropogenic Nest Substrates by Crested Caracaras." *Southeastern Naturalist* 14:N10–15.

7. Estes, W. A., and R. W. Mannan. 2003. "Feeding Behavior of Cooper's Hawks at Urban and Rural Nests in Southeastern Arizona." *Condor* 105:107–16.

8. Caballero, I. C., J. M. Bates, M. Hennen, and M. V. Ashley. 2016. "Sex in the City: Breeding Behavior of Urban Peregrine Falcons in the Midwestern US." *PLoS ONE* 11: e0159054. Accessed July 27, 2017. http://dx.doi.org/10.1371/journal.pone.0159054.

9. Hindmarch, S., and J. E. Elliott. 2015. "When Owls Go to Town: The Diet of Urban Barred Owls." *Journal of Raptor Research* 49:66–74.

10. Bell, D. A., D. P. Gregoire, and B. J. Walton. 1999. "Bridge Use by Peregrine Falcons in the San Francisco Bay Area." In *Raptors in Human Landscapes: Adaptations to Built and Cultivated Environments*, edited by D. Bird, D. Varland, and J. J. Negro, 15–24. San Diego: Academic Press.

11. Boggie, M. A., R. W. Mannan, and C. Wissler. 2015. "Perennial Pair Bonds in an Accipiter: A Behavioral Response to an Urbanized Landscape?" *Journal of Raptor Research* 49:458–70.

12. Tella, J. L., F. Hiraldo, J. A. Donázar-Sancho, and J. J. Negro. 1996. "Costs and Benefits of Urban Nesting in the Lesser Kestrel." In *Raptors in Human Landscapes: Adaptations to Built and Cultivated Environments*, edited by D. Bird, D. Varland, and J. J. Negro, 53–60. San Diego: Academic Press.

13. Lin, W.-L., S.-M. Lin, J.-W. Lin, Y. Wang, and H.-Y. Tseng. 2015. "Breeding Performance of Crested Goshawk *Accipiter trivirgatus* in Urban and Rural Environments of Taiwan." *Bird Study* 62:177–84.

14. Boal, C. W., and R. W. Mannan. 1999. "Comparative Breeding Ecology of Cooper's Hawks in Urban and Exurban Areas of Southeastern Arizona." *Journal of Wildlife Management* 63:77–84.

15. Dwyer, J. F., and R. W. Mannan. 2007. "Preventing Raptor Electrocutions in an Urban Environment." *Journal of Raptor Research* 41:259–67.

16. Hindmarch, S., E. A. Krebs, J. Elliott, and D. J. Green. 2014. "Urban Development Reduces Fledging Success of Barn Owls in British Columbia, Canada." *Condor* 116:507–17.

17. Dwyer, J. F. 2009. "Raptor Electrocution: A Case Study on Ecological Traps, Sinks, and Additive Mortality." *Journal of Natural Resources and Life Science Education* 38:93–98.

18. Hager, S. B. 2009. "Human-Related Threats to Urban Raptors." *Journal of Raptor Research* 43:210–26.

19. Morishita, T. Y., A. T. Fullerton, L. J. Lowenstine, I. A. Gardner, and D. L. Brooks. 1998. "Morbidity and Mortality in Free-Living Raptorial Birds of Northern California: A Retrospective Study, 1983–1994." *Journal of Avian Medicine and Surgery* 12:78–81.

20. Hindmarch, S., E. A. Krebs, J. Elliott, and D. J. Green. 2012. "Do Landscape Features Predict the Presence of Barn Owls in a Changing Agricultural Landscape?" *Landscape and Urban Planning* 107:255–62.

21. Gagné, S. A., J. L. Bates, and R. O. Bierregaard. 2015. "The Effects of Road and Landscape Characteristics on the Likelihood of a Barred Owl (*Strix varia*)-Vehicle Collision." *Urban Ecosystems* 18:1007–20.

22. Ramsden, D. J. 2003. "Barn Owls and Major Roads. Results and Recommendations from a 15-Year Research Project." Barn Owl Trust, Ashburton, UK. Accessed July 27, 2017. http://www.barnowltrust.org.uk/wp-content/uploads/Barn_Owls_and_Major_Roads.pdf.

23. Halfwerk, W., L. J. M. Holleman, C. M. Lessells, and H. Slabbekoorn. 2001. "Negative Impact of Traffic Noise on Avian Reproductive Success." *Journal of Applied Ecology* 48:210–19.

24. Merriman, J. W., C. W. Boal, T. L. Bashore, P. J. Zwank, and D. B. Wester. 2003. "Abundance of Diurnal Raptors in Relation to Prairie Dog Colonies: Implications for Bird-Aircraft Strike Hazard." *Journal of Wildlife Management* 71:811–15.

25. Bayne, E. M., C. A. Scobie, and M. Rawson-Clark. 2012. "Factors Influencing the Annual Risk of Bird-Window Collisions at Residential Structures in Alberta, Canada." *Wildlife Research* 39:583–92.

26. Habberfield, M. W., and C. C. St Clair. 2016. "Ultraviolet Lights Do Not Deter Songbirds at Feeders." *Journal of Ornithology* 157:239–48.

27. Murphy, R. K., E. K. Mojica, J. F. Dwyer, M. M. McPherron, G. D. Wright, R. E. Harness, A. K. Pandey, and K. L. Serbousek. 2016. "Crippling and Nocturnal Biases in a Study of Sandhill Crane (*Grus Canadensis*) Collisions with a Transmission Line." *Waterbirds* 39:312–17.

28. Dwyer, J. F., R. E. Harness, B. D. Gerber, M. A. Landon, P. Petersen, D. D. Austin, B. Woodbridge, G. E. Williams, and D. Eccleston. 2016. "Power Pole Density Informs Spatial Prioritization for Mitigating Avian Electrocution." *Journal of Wildlife Management* 80:634–42.

29. Dwyer, J. F. 2006. "Electric Shock Injuries in a Harris's Hawk Population." *Journal of Raptor Research* 40:193–99.

30. Kagan, R. A. 2016. "Electrocution of Raptors on Power Lines: A Review of Necropsy Methods and Findings." *Veterinary Pathology* 53:1030–36.

31. Wendell, D. M., M. J. Sleeman, and G. Kratz. 2002. "Retrospective Study of Morbidity and Mortality of Raptors Admitted To Colorado State University Veterinary Teaching Hospital during 1995 to 1998." *Journal of Wildlife Diseases* 38:101–6.

32. Wiesmüller, T., P. Sömmer, M. Volland, and B. Schlatterer. 2002. "PCDDs/PCDFs, PCBs, and Organochlorine Pesticides in Eggs of Eurasian Sparrowhawks (*Accipiter nisus*), Hobbies (*Falco subbuteo*), and Northern Goshawks (*Accipiter gentilis*) Collected in the Area of Berlin-Brandenburg, Germany." *Archives Environmental Contamination and Toxicology* 42:486–96.

33. Yu, L. H., X. J. Luo, J. P. Wu, L. Y. Liu, J. Song, Q. H. Sun, X. L. Zhang, D. Chen, and B. X. Mai. 2011. "Biomagnification of Higher Brominated PBDE Congeners in an Urban Terrestrial Food Web in North China Based on Field Observation of Prey Deliveries." *Environmental Science and Technology* 45:5125–31.

34. Chen, D., Y. Wang, L. Yu, X. Luo, B. Mai, and S. Li. 2013. "Dechlorane plus Flame Retardant in Terrestrial Raptors from Northern China." *Environmental Pollution* 176:80–86.

35. Elliott, J. E., J. Brogan, S. L. Lee, K. G. Drouillard, and K. H. Elliott. 2015. "PBDEs and Other POPs in Urban Birds of Prey Partly Explained by Trophic Level and Carbon Source." *Science of the Total Environment* 524:157–65.

36. Henny, C. J., and J. E. Elliott. 2007. "Toxicology." In *Raptor Research and Management Techniques.* 2nd ed., edited by D. M. Bird and K. L Bildstein, 351–64. Surrey, BC: Hancock House Publishing.

37. Newton, I., J. A. Bogan, and P. Rothery. 1986. "Trends and Effects of Organochlorine Compounds in Sparrowhawk Eggs." *Journal of Applied Ecology* 23:461–78.

38. Elliott, J. E., S. W. Kennedy, and K. M. Cheng. 2001. "Assessment of Biological Effects of Chlorinated Hydrocarbons in Osprey Chicks." *Environmental Toxicology and Chemistry* 20:866–79.

39. Woodford J. E., W. H. Krasov, M. E. Meyer, and L. Chambers. 1998. "Impact of 2,3,7,8-TCDD Exposure on Survival, Growth, and Behavior of Ospreys Breeding in Wisconsin, USA." *Environmental Toxicology and Chemistry* 17:1323–31.

40. Rattner, B. A., P. C. Mcgowan, N. H. Golden, J. S. Hatfield, P. C. Toschik, R. F. Lukei, Jr., R. C. Hale, I. Schmitz-Alfonso, and C. P. Rice. 2004. "Contaminant Exposure and Reproductive Success of Ospreys (*Pandion haliaetus*) Nesting in Chesapeake Bay Regions of Concern." *Archives Environmental Contamination and Toxicology* 47:126–40.

41. Lindberg, P., U. Sellström, L. Häggberg, and C. A. De Wit. 2004. "Higher Brominated Diphenyl Ethers and Hexabromocyclododecane Found in Eggs of Peregrine Falcons (*Falco peregrinus*) Breeding in Sweden." *Environmental Science and Technology* 38:93–96.

42. Jaspers, V. A., J. Covaci, T. Maervoet, S. Dauwe, S. Voorspoels, P. Schepens, and M. Eens. 2005. "Brominated Flame Retardants and Organochlorine Pollutants in Eggs of Little Owls (*Athene noctua*) from Belgium." *Environmental Pollution* 136:81–88.

43. Danerud, P. O. 2003. "Toxic Effects of Brominated Flame Retardants in Man and Wildlife." *Environmental International* 29:841–53.

44. Ericsson, W., and D. Urban. 2004. "Potential Risks of Nine Rodenticides to Birds and Non-Target Mammals: A Comparative Approach." U. S. Environmental Protection Agency, Washington, DC. Accessed May 24, 2017. http://www.fwspubs.org/doi/suppl/10.3996/052012-JFWM-042/suppl_file/10.3996_052012-jfwm-042.s4.pdf.

45. Albert, C. A., L. K. Wilson, P. Mineau, S. Trudeau, and J. E. Elliott. 2010. "Anticoagulant Rodenticides in Three Owl Species from Western Canada, 1988–2003." *Archives of Environmental Contamination and Toxicology* 58:451–59.

46. Murray, M. 2011. "Anticoagulant Rodenticide Exposure and Toxicosis in Four Species of Birds of Prey Presented to a Wildlife Clinic in Massachusetts, 2006–2010." *Journal of Zoo and Wildlife Medicine* 42:88–97.

47. Rattner, B. A., R. S. Lazarus, J. E. Elliott, R. F. Shore, and N. van den Brink. 2014. "Adverse Outcome Pathway and Risks of Anticoagulant Rodenticides to Predatory Wildlife." *Environmental Science and Technology* 48:8433–45.

48. DeMent, S. H., Jr., J. J. Chisolm, J. C. Barber, and J. D. Strandberg. 1986. "Lead Exposure in an 'Urban' Peregrine Falcon and Its Avian Prey." *Journal of Wildlife Diseases* 22:238–44.

49. García-Fernández, A. J., M. Motas-Guzmán, I. Navas, P. Maria-Mojica, A. Luna, and J. A. Sánchez-García. 1997. "Environmental Exposure and Distribution of Lead in Four Species of Raptors in Southeastern Spain." *Archives of Environmental Contamination and Toxicology* 33:76–82.

50. Tully, T. N., Jr., G. M. Dorrestein, and A. K. Jones. 2009. *Handbook of Avian Medicine.* 2nd ed. Edinburgh, Scotland: Elsevier.

51. Gavier-Widén, D., J. P. Duff., and A. Meredith. 2012. *Infectious Diseases of Wild Mammals and Birds in Europe.* West Sussex, UK: Wiley-Blackwell, John Wiley and Sons, Ltd.

52. Moriguchi, S., M. Onuma, and K. Goka. 2016. "Spatial Assessment of the Potential Risk of Avian Influenza A Virus Infection in Three Raptor Species in Japan." *Journal of Veterinary Medical Science* 78:1107–15.

53. Wrobel, E. R., T. E. Wilcoxen, J. T. Nuzzo, and J. Seitz. 2016. "Seroprevalence of Avian Pox *Mycoplasma gallisepticum* in Raptors in Central Illinois." *Journal of Raptor Research* 50:289–94.

54. Amin, A., I. Bilic, D. Liebhart, and M. Hess. 2014. "Trichomonads in Birds—A Review." *Parasitology* 141:733–47.

55. Boal, C. W., R. W. Mannan, and K. S. Hudelson. 1998. "*Trichomoniasis* in Cooper's Hawks from Arizona." *Journal of Wildlife Diseases* 34:590–93.

56. Kunca, T., P. Smejkalova, and I. Cepicka. 2015. "Trichomonosis in Eurasian Sparrowhawks in the Czech Republic." *Folia Parasitologica* 62:1–5.

57. Strasser, E. H., and J. A. Heath. 2013. "Reproductive Failure of a Human-Tolerant Species, the American Kestrel, Is Associated with Stress and Human Disturbance." *Journal of Applied Ecology* 50:912–19.

58. Hurley, V. G. 2013. "Factors Affecting Breeding Success in the Peregrine Falcon (*Falco peregrinus macropus*) across Victoria 1991–2012." PhD diss., School of Life and Environmental Sciences, Deakin University.

59. Houston, C. S., and F. Scott. 2006. "Entanglement Threatens Ospreys at Saskatchewan Nests." *Journal of Raptor Research* 40:226–28.

60. Martínez, J. E., I. Zuberogoitia, M. V. Jiménez-Franco, S. Mañosa, and J. F. Calvo. 2016. "Spatio-Temporal Variations in Mortality Causes of Two Migratory Forest Raptors in Spain." *European Journal of Wildlife Research* 62:109–18.

61. Arizaga, J., and M. Laso. 2015. "A Quantification of Illegal Hunting of Birds in Gipuzkoa (North of Spain)." *European Journal of Wildlife Research* 61:795–99.

62. Whitfield, D. P., A. H. Fielding, D. R. A. McLeod, and P. F. Haworth. 2004. "The Effects of Persecution on Age of Breeding and Territory Occupation in Golden Eagles in Scotland." *Biological Conservation* 118:249–59.

63. Dawson, J. W., and R. W. Mannan. 1994. "The Ecology of Harris' Hawks in Urban Environments. Arizona Game and Fish Department, Final Report." Urban Heritage Grant LOA G20058-A, Tucson, AZ.

64. Dwyer, J. F., and J. C. Bednarz. 2011. "Harris's Hawk (*Parabuteo unicinctus*)." In *The Birds of North America*, edited by P. G. Rodewald. Ithaca: Cornell Lab of Ornithology. Accessed July 27, 2017. https://birdsna.org/Species-Account/bna/species/hrshaw.

CHAPTER 15

Human-Raptor Conflicts in Urban Settings

Brian E. Washburn

D O YOU RECALL THE FIRST time you saw a raptor up close? I suspect it might have been a notable experience. Perhaps your thoughts were similar to those expressed by grade school children who were with me on a recent banding trip involving ospreys (*Pandion haliaetus*): "Wow! Look at those talons! I would not want to be a fish," "I don't want to get too close; it might bite me," and "The feathers are so colorful, and it has long powerful wings!" Children truly look at new things with wide-eyed wonder.

People exhibit a variety of emotional responses in the presence of raptors, and these reactions can be intense for those living in the suburbs and cities, especially when the encounters are unexpected. For example, several raptor species—such as ospreys, bald eagles (*Haliaeetus leucocephalus*), and barred owls (*Strix varia*)—are often perceived as species that live only in remote wilderness settings. However, in recent decades, these birds have shown a high degree of adaptability and have become increasingly abundant in many urban and suburban landscapes.[1,2,3] The presence of raptors in suburban and urban areas greatly increases the potential for human-raptor interactions, which can be positive or negative depending on the situation. The reactions and emotions expressed are based on an individual's personal experiences, cultural heritage, education, and other factors. For example, owls have important significance among different human cultures and are symbolic of a wide variety of things, including (but not limited to) intelligence, education, wisdom, misfortune, dark omens, sickness, and death.[4] Commonly, emotional responses to the presence of a raptor in a neighborhood or backyard might include curiosity, awe, pleasure, excitement, concern, outright fear, or indifference.

The intensity of emotional responses to urban raptors can be greatly increased during human-raptor conflict situations, especially in urban areas where residents might not often interact with birds of prey. Additionally, the emotional responses and associated perspectives can vary considerably for the same person depending on the specifics of the human-raptor conflict itself. For example, because I am a raptor researcher and wildlife biologist, I tend to be objective and normally focus on the management of raptor populations (figure 15.1). Managing populations, as opposed to individual animals, is a foundational tenet of the wildlife profession.[5] However, when faced with a human-raptor conflict that affects me personally, such as a locally nesting hawk attacking my pets, a family member, or a backyard chicken flock, I am quite sure my reactions would be subjective and focused on managing (or likely, removing) the offending individual raptor(s). Understanding and appreciating the fact that anyone might shift along the objective-subjective spectrum (figure 15.1), due to a variety of reasons, is important.

Benefits

Urban raptors can provide important benefits and environmental services for humans. Raptors are popular with bird watchers and much of the general public.[6,7] Also, urban raptors can provide opportunities for environmental education of people living in the suburbs and cities who might have limited contact with nature.

Populations   *Individuals*

Objective   *Subjective*

Figure 15.1. Graphic representation of the spectrum of human perspectives related to the resolution of human-wildlife conflicts.

Ecosystem Services

Ecosystem services are benefits people receive from ecosystems and the species that compose those systems. Raptors in urban environments can provide a number of ecosystem services to the human inhabitants of suburban and urban environments.[8] Examples of ecosystem services that raptors provide might include, scavenging by vultures and other raptors, biological control of pest species (e.g., rats [*Rattus* spp.], rock pigeons [*Columba livia*], house sparrows [*Passer domesticus*]), and economic and social benefits from bird-watching and ecotourism associated with urban wildlife.[6,7,9]

Environmental Education

Recent research in childhood education demonstrates how important experiences with nature are in shaping early environmental consciousness[10,11] and ultimately the expression of positive environmental attitudes and behaviors during adulthood.[12,13,14] Some species of urban raptors, due to their prominent nest sites and ability to live in close proximity to humans, provide opportunities for observation and learning (figure 15.2). Furthermore, a raptor's ecological role within an ecosystem provides excellent examples for teaching about basic ecological principles, such as interactions among trophic levels, effects of environmental contaminants, animal migration, and conflict resolution, as we shall see in chapter 16.

As one example, the life history of ospreys can serve as a medium by which children and adults can learn about coastal and estuarine ecosystems, wildlife conservation, and bird migration.[15] Currently, most osprey education programs are locally or regionally focused (i.e., they engage students within one school or community), and many involve primarily elementary school–aged children. During the last decade, improvements to technologies used in raptor research and conservation, especially satellite telemetry and webcams, have provided new scientific insights regarding raptors that migrate long distances. These advances in technology, social media, and web-based data sharing present new opportunities for large-scale citizen science endeavors and projects.[16,17,18] Integrating raptor research conducted in the suburbs and cities into environmental education (at the primary, secondary, and postsecondary levels) and public outreach can provide mutual benefits to all involved.

Figure 15.2. Urban nesting raptors, such as ospreys, provide excellent opportunities for environmental education of people and a chance for them to have a very personal, positive experience with nature in their own backyard.

Challenges with Urban Raptors

Urban raptors can negatively affect a variety of human interests, including human health and safety, livestock and companion animals, and important natural resources. Conflicts between raptors and people generally are localized and often site specific. However, the economic and social effects on the individuals involved can be severe.

HUMAN HEALTH AND SAFETY

Raptors can be aggressive toward humans and inflict serious cuts and lacerations with their talons. In particular, during the nesting season, several species of raptors will defend their nests and young with great ferocity, as will be shown in several case studies detailed in chapter 17. These negative interactions occur most frequently when raptors are nesting in urban and suburban areas. Raptor attacks on humans result in highly charged emotional situations for those who are affected, especially when the incidents involve children. Although such situations usually do not involve the general public and thus might not be a major concern to many people, those individuals directly experiencing a human-raptor conflict can be deeply affected. Wildlife managers and others involved in the resolution of these human-wildlife conflicts must be conscientious of this perspective.

In contrast to the occasional situation in which a pair of raptors is aggressive near their nest, birds of prey inhabiting airports located in urban and suburban areas present potential human-raptor conflict on a much larger scale. This is because of the number of individual birds and multiple species that can be involved, the size of the area affected, and the potential number of humans that could be affected. Raptors pose a risk to safe aircraft operations due to collisions between birds and aircraft (also known as bird strikes). Airports and military airfields represent a unique land use, particularly in suburban or highly urbanized environments.[19] Many species of raptors commonly use the large, open grassland-like habitats for foraging. The relatively large body mass of raptors (relative to other bird species) increases the risk of damage to aircraft as well as the potential for human injuries and fatalities (figure 15.3). Bird strikes involving

Figure 15.3. Collisions between large raptors, such as this adult bald eagle, and aircraft can result in serious damage to aircraft and human injuries. Photo by Chris Cooper.

urban-dwelling raptors have resulted in significant and costly damage to aircraft as well as human injuries and fatalities.[20,21,22]

Nest Locations / Human-Made Structures

The network of communication towers (e.g., lattice and monopole cellular telephone towers and lattice-guyed digital television antennas) is growing exponentially across North American landscapes, notably in the suburbs and cities.[23,24] Communication towers, especially cellular towers, seem to have characteristics that make these structures attractive nesting sites for ospreys, eagles, and other raptors (figure 15.4). Conflicts arise when nest materials interfere with the function of the transmitting and receiving equipment on the towers (thus resulting in interruptions of service) or when repairs and maintenance must be completed. Increases in tower abundance and distribution due to increasing demands for utilities and services by a growing human population, combined with the physical characteristics of towers that are attractive to nesting raptors, appear to be creating the potential for current and future human-raptor conflicts.

Livestock and Companion Animals

The practice of urban agriculture has been growing in popularity during recent decades.[25,26] In contrast to traditional livestock, urban livestock are commonly perceived as pets or companions, and their owners are often emotionally attached to them.[26] In urban agriculture situations, most livestock depredation problems involving raptors occur with backyard poultry. Domestic poultry (e.g., chickens [*Gallus gallus domesticus*], turkeys [*Meleagris gallopavo*], ducks [*Anas platyrhynchos domesticus*], and geese [*Anser anser domesticus*]) are particularly vulnerable to raptor predation because they are conspicuous, unwary, and usually concentrated in areas that lack escape cover.

Raptors are highly opportunistic predators. Small dogs, cats, and kittens left outside and unattended might be at risk from attack and predation from urban raptors. The frequency and severity of such incidents may increase during winter when food is scare or during early summer when newly fledged young are developing their hunting skills.

Figure 15.4. Osprey nest on a cell tower. Situations like this result in human-raptor conflicts that can affect both people and birds.

Effects on Other Wildlife Species

Hawks and owls can negatively affect other species of wildlife by predation and additive mortality. Great horned owls (*Bubo virginianus*), and occasionally red-tailed hawks (*Buteo jamaicensis*), can severely impact colonial waterbird and shorebird nesting colonies by concentrating their hunting efforts on specific colony sites and attacking both young and adults. This can be especially problematic if the nesting birds are rare or have threatened and endangered species status. For example, endangered California least terns (*Sternula antillarum browni*) are preyed upon by several species of raptors within their nesting colonies located at beaches near Oceanside, California.[27] Great horned owls are formidable predators that can kill and eat other raptors, such as peregrine falcons (*Falco peregrinus*) and ospreys. Such predation events can be costly to reintroduction programs with the goal of increasing raptor numbers in urban and suburban areas.

Economic Impacts

The economic impact of human-raptor conflicts in urban areas can be substantial. These might include costs related to direct property damage or loss, lost revenue related to the inability to develop property, liability associated with failure to reduce or prevent human health and safety incidents associated with raptors, and other financial consequences. Although the monetary value of the loss of a few backyard chickens might seem insignificant, this could present an important financial hardship to the individual homeowner.

The presence of a bald eagle nest on private or commercial property could limit the potential economic value of the property due to federal and state regulations designed to protect this species (e.g., Bald and Golden Eagle Protection Act of 1940). Raptor-aircraft collisions have resulted in substantial financial losses associated with damaged and destroyed aircraft.[21,22] Personal and corporate liability of individuals or corporations that are not working to resolve human-raptor conflicts represents a very contemporary and important issue.[28]

Human-Raptor Conflict Management

As with any human-wildlife conflict situation, public information and education regarding wildlife damage management, the ecology of the species involved, and consideration and appreciation of human perceptions and social values are essential components of an effective solution.[29] In particular, understanding the biology of the species involved is paramount to ensure that management efforts are successful. This knowledge will facilitate problem resolution, as each human-raptor conflict situation has the potential to be unusual, whereas other conflicts might occur across a wide variety of scales.

Legal Status

All raptors in the United States are federally protected under the Migratory Bird Treaty Act of 1918 (16 USC, 703–711). Raptors are typically protected under state wildlife laws or local ordinances as well. These laws strictly prohibit the capture, killing, or possession of hawks or owls (or their parts) without a special permit (e.g., federal depredation permit) issued by the US Fish and Wildlife Service.

State-issued wildlife damage or depredation permits also may be required to allow actions to alleviate human-raptor conflicts. Permits are not required to frighten or harass depredating migratory birds unless the birds have threatened or endangered species status.

Passive Management

As will be described in chapter 17, passive management efforts typically involve activities that minimize the contact between the offending raptor(s) and the affected people. For example, when an urban raptor becomes aggressive, the behavior is typically associated with nest defense for a specific period during the breeding season. Purposeful efforts to keep people and pets away from raptor nests during this time period can reduce the chance of aggressive hawk attacks and resulting injuries. Placing signs and notices near territorial nesting raptors can be an especially useful means of providing information to others that could help reduce the frequency and severity of negative interactions.[30]

Managing Prey Resources

Backyard poultry depredations and raptor attacks on small companion animals (e.g., pets) represent human-raptor conflicts that occur on a very local scale. Resolution of conflicts on this level often requires modifications to the animal husbandry practices, such as providing appropriate enclosures for the poultry, or supervising pets while they are outdoors.

Airport grasslands often provide habitat for small mammals (e.g., rodents) that could attract raptors to airport environments. Assessing food habits of raptors (e.g., red-tailed hawks) that use airports allows management efforts to be directed toward the specific prey species of concern.[31] Reductions in small mammal populations on airfields (with the intention of reducing forage availability to wildlife) can be accomplished by implementing an integrated pest management program, which might include the use of effective, targeted pesticide applications, habitat management actions, or other tools. Toxic baiting applications (e.g., rodenticides such as zinc phosphide) that target small mammals and reduce prey population abundance might be effective in reducing raptor use of airfield environments.[32,33] Also, vegetation management activities (i.e., mowing) have resulted in reduced small mammal presence within grassland habitats.[34,35]

Habitat Management

Habitat management approaches applied on a local scale might include the removal of a tree limb to prevent an aggressive raptor pair from using the same site during the next breeding season. Commercially available antiperching devices (e.g., Nixalite, Cat Claws, and inverted spikes) might help reduce use of buildings, roofs, and other structures that problematic raptors use as perching and hunting sites.[36] Overhead wires made of nylon cord or heavy monofilament fishing line suspended in parallel over poultry pens and pet runs can also be effective in deterring urban raptors from swooping down on their intended prey.[37]

In situations where urban raptors (e.g., ospreys) are nesting on human-made structures and thus may create conflicts, the addition of an artificial nesting platform (either to the structure itself or preferably on a pole erected nearby the nest site) offers a mutually beneficial solution.[38] Discouraging the birds from using problematic nest sites through modifications to the structures, concurrent with the installation of the artificial platform, is an essential part of the process. Such actions can reduce fires, electrocutions, and collisions with power transmission towers and transmission lines. Decreased perching and nesting near such human structures in urban areas can reduce raptor mortality and decrease the occurrence of human-raptor conflicts.[38,39]

Active Management

There are many tools and techniques that can be used to disperse raptors from an area where they are causing damage to property or conflicts with people. Simple, inexpensive tools and methods (e.g., air horns, banging pots and pans, and other methods of making loud noises at the offending bird) can be effective, especially in the suburbs and cities areas where deployment of pyrotechnics and firearms is not advisable or even legal. The effectiveness of frightening devices depends greatly on the bird, area, season, and method of application. Frightening devices usually reduce rather than totally eliminate human-raptor conflicts, but this might be sufficient for the involved individuals to consider the effort a "success."

Human safety issues related to raptor-aircraft collisions are contemporary and serious. Addressing raptor strike events requires enhanced aviator awareness, proper reporting, and an emphasis on resolving the issue. Reduction of risk posed by raptors (and other wildlife) to aviation safety is best effected through the use of an integrated wildlife damage management program.[40] These programs often include the use of nonlethal hazing and harassment, installation of antiperching devices on airport structures, audible noise deterrents, pyrotechnics, translocation or culling of problematic individuals, and a variety of other tools and techniques.[40,41,42]

Livetrapping and translocating problem raptors is a commonly used, nonlethal method of resolving conflict situations between humans and raptors (figure 15.5). If possible, experienced birdbanders or wildlife professionals with proper training should manage raptor livetrapping efforts to ensure the safety of both raptors and people. As with other activities associated with raptors (such as rehabilitation[42]), state and federal permits are required to conduct livetrapping and translocation activities.

Translocating raptors is often perceived as a "panacea" for resolving conflicts. The homing behavior of raptors can be strong, and the offending bird(s) might return and continue the conflict. Furthermore, the efficacy of raptor translocation is relatively unstudied.[43,44] Additional well-designed and peer-reviewed research is needed to allow for the development of effective, science-based recommendations to resolve these problems.

Figure 15.5. A young red-tailed hawk caught near a runway at an airport in a highly urbanized landscape. Live-capture and translocation of urban raptors away from conflict situations is a nonlethal method of conflict resolution.

Under the most extreme of circumstances, the lethal removal of problem urban raptors might be the most appropriate solution to human-raptor conflicts. Although this might be the initial desire of individuals experiencing high-stress conflict situations (e.g., raptor attacks on children that cause injuries), the application of this method is complex and requires state and federal permits. Lethal removal is usually considered only when all other conflict-reduction methods have been exhausted.

Summary

Interactions between raptors and humans in urban environments can be beneficial or detrimental, depending on a number of factors. Urban raptors provide environmental and social benefits to city-dwelling humans. However, urban raptors can also be involved in a wide variety of human-raptor conflict situations, some of which involve human health and safety. A multitude of factors and considerations can influence the decision on how to manage human-raptor conflicts. One of the most important questions to ask is "What is success?" Answering this question should involve those directly affected by the conflict situation. Overall, integrated management approaches are the most effective for long-term solutions. Management actions must be science based but also should be sensitive to human needs and desires.

Literature Cited

1. Poole, A. F., R. O. Bierregaard, and M. S. Martell. 2002. "Osprey (*Pandion haliaetus*)." In *The Birds of North America*, no. 683, edited by A. Poole and F. Gill. Philadelphia: The Academy of Natural Sciences; Washington, DC: American Ornithologists' Union.

2. Rutz, C. 2008. "The Establishment of an Urban Bird Population." *Journal of Animal Ecology* 77:1008–19.

3. Hager, S. B. 2009. "Human-Related Threats to Urban Raptors." *Journal of Raptor Research* 43:210–26.

4. Duncan, J. R. 2003. *Owls of the World: Their Lives, Behavior, and Survival.* Buffalo: Firefly Books Press.

5. Bolen, E. G., and W. L. Robinson. 1999. *Wildlife Ecology and Management.* 4th ed. Upper Saddle River, NJ: Prentice-Hall.

6. Galbraith, J. A., J. R. Beggs, D. N. Jones, E. J. McNaughton, C. R. Krull, and M. C. Stanley. 2014. "Risks and Drivers of Wild Bird Feeding in Urban Areas of New Zealand." *Biological Conservation* 180:64–74.

7. Donázar, J. A., A. Cortés-Avizanda, J. A. Fargallo, A. Margalida, M. Moleón, Z. Morales-Reyes, R. Moreno-Opo, J. M. Pérez-García, J. A. Sánchez-Zapata, I. Zuberogoitia, and D. Serrano. 2016. "Roles of Raptors in a Changing World: From Flagships to Providers of Key Ecosystem Services." *Ardeola* 63:181–234.

8. Daily, G. C., S. Polasky, J. Goldstein, P. M. Kareiva, H. A. Mooney, L. Pejchar, T. H. Ricketts, J. Salzman, and R. Shallenberger. 2009. "Ecosystem Services in Decision Making: Time to Deliver." *Frontiers in Ecology and the Environment* 7:21–28.

9. Whelan, C. J., Ç. H. Şekercioğlu, and D. G. Wenny. 2015. "Why Birds Matter: From Economic Ornithology to Ecosystem Services." *Journal of Ornithology* 156:227–38.

10. Hinds, J., and P. Sparks. 2008. "Engaging with the Natural Environment: The Role of Affective Connection and Identity." *Journal of Environmental Psychology* 28:109–20.

11. Cheng, J. C., and M. C. Monroe. 2012. "Connection to Nature: Children's Affective Attitude toward Nature." *Environment and Behavior* 44:31–49.

12. Wells, N., and K. Lekies. 2006. "Nature and the Life Course: Pathways from Childhood Nature Experiences to Adult Environmentalism." *Children, Youth and Environments* 16:2–25.

13. Chawla, L., and D. F. Cushing. 2007. "Education for Strategic Environmental Behavior." *Environmental Education Research* 13:437–52.

14. Collado, S., H. Staats, and J. A. Corraliza. 2013. "Experiencing Nature in Children's Summer Camps: Affective Cognitive and Behavioural Consequences." *Journal of Environmental Psychology* 33:37–44.

15. Cushing, R., and B. E. Washburn. 2014. "Exploring the Role of Ospreys in Education." *Journal of Raptor Research* 48:414–21.

16. Devictor, V., R. J. Whittaker, and C. Beltrame. 2010. "Beyond Scarcity: Citizen Science Programmes as Useful Tools for Conservation Biogeography." *Diversity and Distributions* 16:354–62.

17. Dickinson, J. L., B. Zuckerberg, and D. N. Bonter. 2010. "Citizen Science as an Ecological Research Tool: Challenges and Benefits." *Annual Review of Ecology, Evolution, and Systematics* 41:149–72.

18. Catlin-Groves, C. L. 2012. "The Citizen Science Landscape: From Volunteers to Citizen Sensors and Beyond." *International Journal of Zoology* 12:1–14.

19. DeVault, T. L., B. F. Blackwell, and J. L. Belant, eds. 2013. *Wildlife in Airport Environments: Preventing Animal-Aircraft Collisions through Science-Based Management.* Bethesda: Johns Hopkins University Press.

20. Washburn, B. E. 2014. "Human–Osprey Conflicts: Industry, Utilities, Communication, and Transportation." *Journal of Raptor Research* 48:387–95.

21. Washburn, B. E., M. J. Begier, and S. E. Wright. 2015. "Collisions between Eagles and Aircraft: An Increasing Problem in the Airport Environment." *Journal of Raptor Research* 19:192–200.

22. Dolbeer, R. A., J. R. Weller, A. L. Anderson, and M. J. Begier. 2016. "Wildlife Strikes to Civil Aircraft in the United States 1990–2015." Serial Report Number 22. Federal Aviation Administration, National Wildlife Strike Database, Washington, DC.

23. Manville, A. M., II. 2005. "Bird Strikes and Electrocutions at Power Lines, Communication Towers, and Wind Turbines: State of the Art and State of the Science—Next Steps toward Mitigation." In *Bird Conservation Implementation in the Americas: Proceedings of the 3rd International Partners in Flight Conference*, edited by C. J. Ralph and T. D. Rich, 1051–64. USDA Forest Service, Gen. Tech. Rep. PSW-GTR-191. Albany, CA: Pacific Southwest Research Station.

24. Manville, A. M., II, 2009. "Towers, Turbines, Power Lines, and Buildings: Steps Being Taken by the U.S. Fish and Wildlife Service to Avoid or Minimize Take of Migratory Birds at These Structures." In *Tundra to Tropics: Connecting Habitats and People*, edited by T. D. Rich, C. Arizmendi, D. Demarest, and C. Thompson, 262–72. McAllen, TX: Proceedings of the 4th International Partners in Flight Conference.

25. De Zeeuw, H., R. Van Veenhuizen, and M. Dubbeling. 2011. "The Role of Urban Agriculture in Building Resilient Cities in Developing Countries." *Journal of Agricultural Sciences* 149:153–63.

26. Huang, D., and M. Drescher. 2015. "Urban Crops and Livestock: The Experiences, Challenges, and Opportunities of Planning for Urban Agriculture in Two Canadian Provinces." *Land Use Policy* 43:1–14.

27. Butchko, P. H., and M. A. Small. 1992. "Developing a Strategy of Predator Control for the Protection of the California Least Tern: A Case History." *Proceedings of the Vertebrate Pest Conference* 15:29–31.

28. Dale, L. A. 2009. "Personal and Corporate Liability in the Aftermath of Bird Strikes: A Costly Consideration." *Human–Wildlife Conflicts* 3:216–25.

29. Conover, M. 2002. *Resolving Human–Wildlife Conflicts: The Science of Wildlife Damage Management.* Boca Raton, FL: CRC Press.

30. See chapter 17.

31. Washburn, B. E., G. E. Bernhardt, and L. A. Kutschbach-Brohl. 2011. "Using Dietary Analysis to Reduce the Risk of Wildlife–Aircraft Collisions." *Human–Wildlife Interactions* 5:204–9.

32. Witmer, G. W., and J. W. Fantinato. 2003. "Management of Rodent Populations at Airports." *Proceedings of the Wildlife Damage Management Conference* 10:350–58.

33. Witmer, G. W., R. Sayler, D. Huggins, and J. Capelli. 2007. "Ecology and Management of Rodents in No-Till Agriculture in Washington, USA." *Integrative Zoology* 2:154–64.

34. Seamans, T. E., S. C. Barras, G. E. Bernhardt, B. F. Blackwell, and J. D. Cepek. 2007. "Comparison of 2 Vegetation-Height Management Practices for Wildlife Control at Airports." *Human–Wildlife Conflicts* 1:97–105.

35. Washburn, B. E., and T. W. Seamans. 2007. "Wildlife Responses to Vegetation Height Management in Cool-Season Grasslands." *Rangeland Ecology and Management* 60:319–23.

36. Washburn, B. E. 2016. "Hawks and Owls." *Wildlife Damage Management Technical Series*. Fort Collins, CO: USDA, Animal and Plant Health Inspection Service (APHIS), Wildlife Services National Wildlife Research Center.

37. Marsh, R. E., W. A. Erickson, and T. P. Salmon. 1991. "Bird Hazing and Frightening Methods and Techniques." California Department of Water Resources, Contract Number B-57211, Davis, CA.

38. Avian Power Line Interaction Committee (APLIC). 2012. "Reducing Avian Collisions with Power Lines: The State of the Art In 2012." Edison Electric Institute and APLIC, Washington, DC.

39. Harness, R. E., and K. R. Wilson. 2001. "Electric-Utility Structures Associated with Raptor Electrocutions in Rural Areas." *Wildlife Society Bulletin* 29:612–23.

40. US Department of Agriculture (USDA). 2005. "Managing Wildlife Hazards at Airports." USDA Animal and Plant Health Inspection Service, Wildlife Services, Washington, DC.

41. Olexa, T. J. 2006. "An Integrated Management Approach for Nesting Osprey to Protect Human Safety and Aircraft at Langley AFB, Virginia." *Proceedings of the Vertebrate Pest Conference* 22:216–21.

42. See chapter 16.

43. Guerrant, T. L., C. K. Pullins, S. F. Beckerman, and B. E. Washburn. 2013. "Managing Raptors to Reduce Wildlife Strikes at Chicago's O'Hare International Airport." *Proceedings of Wildlife Damage Management Conference* 15:63–68.

44. Schafer, L. M., and B. E. Washburn. 2016. "Managing Raptor-Aircraft Collisions on a Grand Scale: Summary of a Wildlife Services Raptor Relocation Program." *Proceedings of the Vertebrate Pest Conference* 27:248–52.

CHAPTER 16

Raptors as Victims and Ambassadors: Raptor Rehabilitation, Education, and Outreach

Lori R. Arent, Michelle Willette, and Gail Buhl

F ROM HUNTING SONGBIRDS AT BACKYARD feeders to circling in awe-inspiring loops above a city park, raptors make their presence known in urban environments. As raptors increasingly inhabit urbanized areas, they not only become more visible to people but encounter new opportunities and challenges (see chapter 14). Challenges can include interactions with humans or elements of the landscape that result in raptors being injured, being poisoned, contracting diseases, or in some cases, experiencing unnecessary interruptions to normal life processes by well-meaning people. But raptors interacting with humans can also increase awareness of and appreciation for the birds, fueling community interest and intensifying fascination and compassion for these charismatic apex predators, as we shall see in chapter 17. As a result, the fields of raptor rehabilitation, professional education in veterinary and rehabilitation science, and public outreach have grown. These fields are interrelated, and each one is complex, encompassing special qualifications, engagement of others, ethical considerations, and much more.

Raptor Rehabilitation

More than forty years ago, wildlife rehabilitation became a field of interest for people concerned with the welfare and conservation of native wildlife. Two organizations were established to create standards of care for injured and orphaned wild animals and to collect and disseminate information: the International Wildlife Rehabilitation Council (IWRC) in California in 1972 and the National Wildlife Rehabilitator's Association (NWRA) in Minnesota in 1982.[1,2] Both of these organizations are still active today and, collectively, have accrued more than 2,500 members worldwide. Smaller state organizations formed along with the national groups to help promote learning and networking between members, birding organizations, researchers, and the general public.

With the vast amount of medical, behavioral, and biological knowledge required for wildlife rehabilitation to be conducted professionally and ethically, people often specialize. One specialty, raptor rehabilitation, provides medical and physiotherapy services to injured and orphaned birds of prey with the goal of reintegrating them into the wild population. To engage in this activity, rehabilitators must acquire both state and federal migratory bird possession permits, as indicated in the Code of Federal Regulations (CFR; 50 CFR 21.31).[3]

Raptor rehabilitation permits are accompanied by specific reporting requirements that must be met for permit renewal. Within 24 hours of receiving a bald or golden eagle, or any species listed as threatened or endangered under the Endangered Species Act (50 CFR 17),[3] rehabilitators must notify state and federal authorities. Rehabilitators must also notify state and federal law enforcement offices within 48 hours of receiving a patient that is confirmed to have been shot, trapped (using lethal traps set for other animals), electrocuted or shocked, or poisoned; a few states require the reporting of other anthropogenic causes of injury in this time frame as well. In addition to these real-time notifications, rehabilitators who treat raptors and other migratory bird species must submit an annual report to both state and federal agencies for permit renewal. A state report is created by each state. The reporting requirements for state reports vary widely and often are not as comprehensive as the standardized federal report. The federal report (Service Form 3-202-4) was designed by the US Fish and Wildlife Service (FWS) and includes information such as total number of raptors admitted by species and the resolutions, cases pending, cases transferred to other permit holders, and reportable injuries (projectiles, toxicity, electrocution, trapping). Most recently, an optional section was added requesting information on confirmed cases of disease or toxicity.

Migratory bird rehabilitation permits also come with a list of additional conditions, two of which are particularly noteworthy. First, migratory birds can only be held for a maximum of 180 days for rehabilitation, after which special permission must be granted for continuance of care. Second, birds that are undergoing rehabilitation cannot be put on public display.

For the rehabilitation of migratory birds, each of the 50 states is under the jurisdiction of one of eight FWS regions; this division, however, is where consistency ends. States have the authority to be more, but not less, restrictive than federal agencies and, therefore, often set their own permit conditions. Some states grant an umbrella permit that allows all employees and volunteers of an established organization to assist in the rehabilitation process at a designated location. Other states require each person to possess his or her own permit. A few states have a tiered permitting structure based on the level of knowledge and experience a permit holder demonstrates (novice, general, and master). For long-term care, raptors require at least a general permit status in this structure. Other states refuse to grant raptor rehabilitation permits to individuals entirely.

Federal migratory bird possession permits are required of all rehabilitators treating migratory birds, not just raptors. In 2015, the FWS issued 1,369 migratory bird possession permits for rehabilitation of all migratory birds nationally, but they are not analyzed by bird type (L. Harrison, pers. comm.). Thus collecting accurate information on the number of people specifically involved in raptor rehabilitation or the number and species of raptors treated is extremely challenging.

To help create a picture of raptor rehabilitation in at least one federal region, the Raptor Center (TRC) at the College of Veterinary Medicine at the University of Minnesota performed a descriptive analysis of all 2011 rehabilitation annual reports submitted for region 3 (upper Midwest). Analysis of these reports from 216 permittees showed that more than 45,000 migratory birds were admitted for rehabilitation; almost one-third of the permittees saw 10 birds or fewer per year, and only eight permittees admitted more than 40 percent of all the birds. The best estimate is that in this region, approximately 7,400 raptors were admitted to rehabilitation facilities.

The NWRA and IWRC directories list more than 62 institutions with the descriptors of raptor, bird of prey, owl, and so on in their institutional name, which would seem to signify a specialty, or exclusivity, in raptors. Of the 1,343 US members of NWRA, 608 indicated specialties; of these self-described members, 242 list raptors as a specialty. Some of these are 501(c)(3) nonprofit organizations that are financially supported by philanthropy. In-kind contributions, such as volunteer services and supplies, are often critical to the daily operations of these centers. Individuals who are permitted to conduct raptor rehabilitation, many of whom admit relatively few birds per year, often personally absorb the majority of the costs incurred.

Most urban raptors brought to rehabilitation facilities are there for anthropogenic reasons; that is, they are victims of the human-animal interface.[1] Many suffer from traumatic injuries resulting from the following:

- collisions with moving objects (e.g., cars, trains, planes)
- collisions with stationary objects (e.g., windows, buildings, power lines)
- entrapment (e.g., in leg-hold traps, spent fishing line, recreational nets such as soccer nets, holiday light strings, barbed wire fences, chimneys, buildings, or oil)
- projectiles (e.g., shotgun, rifle, and BB gun pellets or arrows)
- interactions with domestic animals

Birds may suffer from one or multiple types of injuries from these ordeals. Fractures, soft tissue trauma, head trauma, neurological disorders (spinal trauma, nerve damage, traumatic brain injury), internal trauma, and significant damage to flight and tail feathers are common clinical findings upon admission. Due to the length of time between incident and rescue, starvation is often a secondary ailment. Other anthropogenic causes of admission (see figure 16.1) include intoxication from environmental contaminants like lead, organophosphates, or rodenticides; habitat destruction such as clear-cutting and/or nest tree removal during the nesting season, which may result in orphaned nestlings; and unnecessary human intervention (e.g., people disturbing young fledglings incapable of full flight when they see no parents in the vicinity). In addition, raptors are admitted due to infectious disease, such as West Nile virus, and *failure to thrive* (a term most often used to describe juveniles that have difficulty maintaining weight once parental care is no longer provided). These categories are similar to those reported by others, with the exact percentages dependent on geographical location and species.[4,5]

The primary reasons people get involved in raptor rehabilitation are to relieve animal suffering and to return injured and orphaned birds to the wild. Best practices are to release only those individuals that are physically and mentally sound, with the possibility that some of them will breed to perpetuate the population. If a raptor is unable to perform survival activities such as flying in its characteristic

Figure 16.1. Common causes of admission to urban
rehabilitation facilities. *A*, collision with moving vehicle;
B, entrapment by building structure (chimneys, chimney
caps); *C*, habitat destruction.

style (e.g., fast-flapping flight, soaring, gliding, hovering, stooping), hunting successfully, avoiding other predators, and traveling hundreds to thousands of miles if migratory, it should not be released. Rehabilitators evaluate each individual and consider its age, hunting strategy, migration status, and the time of year in the decision-making process. They use prerelease preparations in the form of physical reconditioning and live-prey testing to help evaluate a raptor's physical abilities and provide insight into its ability to meet the challenges of life in the wild. Release rates vary widely between institutions due to the lack of standardization in many aspects of raptor rehabilitation (K. MacAulay, pers. comm.).

Following sound ecological principles, those raptors deemed releasable should most often be released at the location of recovery (i.e., where the bird was captured). However, exceptions often include the following:

- releases of raptors whose recovery location is unknown
- birds in midmigration
- adult birds whose territories are known to be occupied
- birds admitted due to direct loss of habitat
- birds recovered near airports (or other areas with public safety concerns)
- birds recovered from "unsafe" locations or circumstances

In urban landscapes, the question often becomes, What is the definition of *safe*? It is obviously not safe to return a recently fledged American kestrel to the middle of the busy downtown area where it originated or to return a great horned owl recovered from a backyard chicken coop to that location. However, if rehabilitators choose alternative release locations, they may not know how the presence of one additional raptor will affect the balance of the ecosystem. Are the birds any "safer" in the new location, or are they just exposed to different dangers (predation from other species, territorial battles, etc.)?

In addition to release, there are two other potential outcomes: euthanasia and placement at permitted facilities. Birds suffering from severe injuries are humanely euthanized; others that are deemed nonreleasable either at the time of admission or following the rehabilitation process are evaluated for possible placement at permitted educational facilities.[6] However, throughout the rehabilitation community, there are differences of opinion on criteria for release, immediate euthanasia, and captive placement. With the increasing emphasis on animal welfare, best practices must be part of the decision-making process for all outcomes. With more than 40 years of experience, TRC has established general guidelines for the different case resolutions (table 16.1).

Table 16.1. The Raptor Center's general guidelines for patient outcomes.

Euthanasia	Release	Placement at educational facility
Nonrepairable wing or leg fractures (includes those requiring amputation)	Full recovery from illness or injury with no physical, visual or radiographic signs of complications	Injury will not lead to future complications (structural integrity) or chronic pain
Nonrepairable trauma to tendons or ligaments affecting function		Injury will not predispose to bumble foot (pathology of the avian foot, often a result of uneven weight bearing on both feet)
Significant joint trauma	Goals of physical reconditioning program met (normal flight mechanics, strength, endurance for species)	Injury will not prevent opportunities to thrive (perch, move between two perches, self-feed, bathe, and conduct species-specific behaviors)
Spinal trauma with posterior paralysis		
Bilateral blindness	Behavioral assessment passed (imprint status, live prey acquisition)	Behavioral assessment passed (demonstration of a calm, nonaggressive and nonself-destructive temperament in captivity)
Toxicities with intractable clinical signs		
Starvation as defined by specific physiologic values	A complete set of intact remiges and rectrices	Blood values fall within an acceptable range
Nonsalvageable injuries or incurable diseases	Blood values fall within an acceptable range	Appropriate placement available. Facility authorized; knowledgeable about management, training, and welfare standards (offering opportunities for choice and control); has access to veterinary services

One of the purported opportunities of raptor rehabilitation is access to birds and the health data associated with them. As apex predators, raptors are sentinels of environmental health.[7] They demonstrate the centuries-old "One Health" concept—that the health of humans, animals, and the environment are interrelated.[7,8] Given the nature of raptor rehabilitation, much of the information collected is in the realm of citizen science, and many wildlife care centers have published articles containing admission and outcome data. Usually these data are derived from annual reports submitted to regulatory authorities and suffer from multiple issues including incomplete, inconsistent, and inaccurate data; poor compliance with instructions; and differing interpretations of the instructions.

As the human population continues to increase, there is a growing need for the collation of wildlife and environmental health data to help inform public policy in human, animal, and ecosystem health. The quantity and quality of these rehabilitation data could be vastly improved by unifying reporting requirements across regulatory agencies, providing more guidance and training on the requested information, and using technology to standardize data entry. Data aggregated from multiple centers over a range of years could then be compiled into meaningful information to inform public policy decisions. This information could include long-term population monitoring of the spatial and seasonal distribution of species, the presence of environmental toxins, the occurrence of invasive species, and the monitoring of emerging and zoonotic diseases. The lack of a comprehensive, integrated database impedes the collection of these wildlife and environmental health data.

To develop the infrastructure, data tools, and liaisons needed to use the information from animals presented for rehabilitation, TRC formed the Clinical Wildlife Health Initiative (CWHI) in 2011.[9] The CWHI has worked with a cross section of wildlife rehabilitation professionals and wildlife care centers to create, share, and steward standardized terminology unique to wildlife rehabilitation. The CWHI also works with developers of medical records programs (e.g., WILD-ONe, Raptor Med, and Wildlife Rehabilitation MD) to encourage the use of this standardized terminology in their programs and has approached regulatory authorities about incorporating this standardized data set and terminology into administrative reporting forms.

The field of clinical wildlife medicine is also growing. Numerous raptor rehabilitation centers are now staffed by veterinarians, and many are associated with universities or colleges of veterinary medicine. Raptor medicine includes the following areas of clinical research:

- orthopedics
- ophthalmology
- nonsteroidal anti-inflammatories and other medications used in human and veterinary medicine
- reconditioning physiology
- disease (e.g., trichomoniasis; aspergillosis; West Nile virus, avian influenza, chlamydiosis, and other zoonotic diseases)
- toxicity (e.g., lead poisoning, barbituate poisoning, and effects of second-generation anticoagulant rodenticide)
- electrocution/electric shock
- effects of projectiles (e.g., shotgun, rifle, and BB gun pellets or arrows)

Some of this clinical research has been extended to population-level research and has influenced public policy and law in the areas of endangered and invasive species; hunting, trapping, and fishing practices; power and wind energy; pesticides, rodenticides, and other toxins; and public health and emerging diseases.[7,10] However, there is still much to learn, especially regarding clinical raptor medicine, raptor rehabilitation science, animal welfare and ethical considerations, postrelease monitoring, effects of rehabilitated individuals on population fitness, and effects of rehabilitation on overall populations and ecosystems.

Monitoring survival and activities of raptors following their release can contribute to the body of knowledge on the effectiveness of rehabilitation techniques, causes of morbidity and mortality, nestling dispersal, longevity, migration pathways, and the long-term effects of toxins. It can also help identify potential threats to the ecosystem. Rehabilitators currently employ two postrelease monitoring methods. First, several rehabilitation organizations are authorized to apply aluminum leg bands to raptors for postrelease monitoring. They either possess a United States Geological Survey (USGS) master birdbanding permit or are subpermittees under a master bander. According to the USGS Bird Banding Laboratory, only 200 banded rehabilitated raptors were in the database during the 1960s; currently there are more than 43,000, with the highest number coming from locations with the highest human population.[5] In urban areas, 6–8 percent of banded, rehabilitated, and released raptors are reencountered dead or alive (L. Arent, unpubl. data; D. Scott, pers. comm.).[11]

Two particularly interesting encounters of rehabilitated birds banded by TRC involved a Cooper's hawk (*Accipiter cooperii*) and an osprey (*Pandion haliaetus*). The young Cooper's hawk was admitted in early July 2006 suffering from trichimonas and weight loss. It was treated and reunited with its family later that month and recovered in January 2009 in Texas, more than 1,600 km from its release location. The osprey was admitted in September 2013 with a fractured coracoid (collarbone) and five weeks later was released where she was found. She reportedly built a nest 24 km from her release location in May 2015, and although that nesting effort failed, she successfully fledged young in 2016 and 2017.

Telemetry is the second tool rehabilitators may use to gather information on released individuals. Very High Frequency (VHF) telemetry and satellite-based systems (including Doppler, Global Positioning System [GPS], and Global System for Mobile communication [GSM] technologies) are available but require amendments to rehabilitation permits and are often prohibitively expensive.

Professional Education

An additional facet of raptor rehabilitation is the opportunity to disseminate information. As previously mentioned, several colleges and universities are associated with raptor rehabilitation programs, providing opportunities for veterinary research and wide promulgation of information as well. New methodologies and best practices have become incorporated into established protocols for the medical, surgical, and physiotherapy components of rehabilitation. Our recent literature review found 1,149 articles from 149 different peer-reviewed journals in dozens of scientific disciplines that were based on animals presented for wildlife rehabilitation. Accipitriformes and Falconiformes were the topics of 39 percent and Strigiformes were the topics of 9 percent of the avian-specific articles.

Urban facilities are focal points for providing professional educational opportunities due to their relatively high caseloads. Training programs and workshops for rehabilitators and veterinary professionals are available, and with new technologies, distance learning is providing avenues for training on an international scale. Annual conferences held by national and state organizations promote continuing education, networking, and increased professionalism of the field. In addition, externships, postgraduate certificates, internships, residencies, and board certifications offered by the American Association of Zoo Veterinarians[12] and the American Board of Veterinary Practitioners[13] are available for veterinarians whose career goals focus on wildlife.

Outreach

Information regarding raptors in urban settings is shared with the public both informally upon request and formally through professional education programming. On a daily basis, rehabilitators are called upon to share knowledge with the general public in person or via phone, e-mail, and social media. These timely responses can offer relief to state nongame wildlife agencies, many of which are not sufficiently staffed to handle the large volume of raptor-related inquiries in urban areas.

People possess diverse attitudes toward birds of prey, and rehabilitators are often their first point of contact. Rehabilitators are often called upon to answer questions regarding species' natural history and behavior; safety of adults, children, and pets; nest-site aggression toward people (e.g., by broad-winged hawks [*Buteo platypterus*], red-shouldered hawks [*Buteo lineatus*], peregrine falcons [*Falco peregrinus*]); and unfavorable situations that may be seen when watching online nest cameras. Residents may even demand hawks be removed from their yards due to raiding bird feeders or chicken coops. However, rehabilitation permits are restrictive and do not allow holders to remove birds from nests, move active nests, or trap and remove an unwanted individual from their territory. These types of inquiries must be handled by state and federal authorities and require issuance of special permits (see chapter 17).

Raptor rehabilitators are not the only ones providing information in this fashion to the general public. Environmental educators are increasingly called upon to interpret urban raptor behavior. Many of the same questions are asked, and all individuals providing information need depth and breadth of accurate knowledge about raptor behavior. They need to be able to speak to the public truthfully about the perceived and real conflicts between people and raptors (see chapters 15 and 17). Providing accurate natural science and history information to the public is crucial to helping it make informed choices, especially ones that may affect other living species. It is incumbent on those who present themselves as authorities or educators to have a grasp of the facts and take a balanced approach to instruction. According to *Excellence in Environmental Education: Guidelines for Learning*, "Educators should incorporate differing perspectives and points of view evenhandedly and respectfully, and present information fairly and accurately."[14] Live animal presentations should be done ethically, compassionately, and truthfully. Anything less betrays the animals being represented.[15]

Formal Programming

With the number of human-raptor interactions continuing to increase as rural landscapes are redesigned into urban landscapes, formal educational programming is critical for peaceful coexistence. Several effective methods are used to provide environmental education curricula across the United States, and one of the most common tools is the presentation of live raptors. Most birds are used in programs at nature centers, rehabilitation facilities that have an educational component, environmental learning centers, and zoos. The raptors may be on display, trained to perch on a gloved hand, or trained for flight demonstrations. Education raptors may also be taken off facility grounds for teaching at a variety of other venues. Raptors obtained for educational programming are frequently permanently disabled birds that are transferred from rehabilitation facilities, but physically healthy captive-bred birds and retired falconry birds are also acquired for this purpose.

Maintaining a collection of live native raptors for educational use requires specific federal and state permits (50 CFR 10, 13, 21, 27).[1,16] In 2015, 877 live raptor special purpose possession permits were issued or renewed by the FWS (L. Harrison, pers. comm.). This statistic, however, represents only a portion of the raptors currently employed for educational purposes. Facilities accredited by the Association of Zoos and Aquariums (AZA), public museums with live collections, and public scientific or educational institutions are exempt from federal permit possession (50 CFR part 21.12).[1,16] Many zoos and private educational facilities also use nonnative raptor species for teaching purposes; nonnative raptors are not subject to permits or permit regulations under the Migratory Bird Treaty Act.

In addition to a special purpose possession permit, people exhibiting bald and golden eagles must have an eagle exhibition permit for educational programming under the Bald and Golden Eagle Protection Act (50 CFR 10, 12, 22.21).[1,17] This act covers not only live possession but exhibition of carcasses, body parts, nests, or eggs as well. For noneagles, these educational tools necessitate yet another permit, a salvage permit, which may be required by both state and federal agencies.

Federal education permits come with a few specific conditions. First, since raptors are considered public resources, FWS permit regulations require a minimum of 12 public programs per year per raptor; if a bird is on display only, the facility housing it must be open to the public a minimum of 400 hours annually for programming or tours. Second, program content must focus on wildlife conservation messaging but may also include information on Native American culture, history, and the sport of falconry. Third, the general public is prohibited from coming in direct contact with raptors held for program use. Finally, education raptors cannot be used for commercial purposes.[1,17]

Most states also require a special purpose possession education permit to house and use native live raptors for display or in formal programming. In these states, federal permits are invalid unless a state permit is also issued.

In order to comply with permit stipulations at both the state and federal levels, activities conducted with each raptor held must be reported yearly. The Raptor Center of St. Paul, Minnesota, surveyed the 62 named "raptor" institutions in the NWRA and IWRC directories for their use of raptors as educational ambassadors for conservation messaging. Of the 27 facilities that responded, 81 percent maintained various species of native raptors and had collections ranging from 3 to 93 individuals. Survey respondents reported that the number of people reached at least once annually with a conservation message using a live raptor was approximately 250,000 (K. MacAulay, pers. comm.). The Raptor Center alone presented approximately 1,000 programs in 2015, reaching more than 100,000 people.

Many environmental education curricula are based on professional teaching standards and focus on STEAM concepts (science, technology, engineering, arts, and math). Environmental educators work toward effecting long-term behavioral change by using raptors to put a "face" to conservation challenges. The number of people exposed to live raptors in education programs may be large, but the number of people significantly moved to action is a much smaller number and difficult to quantify. Awareness is an important aspect of environmental education, but a critical component is to prompt positive action toward raptors and the environment. Positive behavioral change in favor of conservation is notoriously difficult to measure in the field of environmental education.[18]

As in the field of raptor rehabilitation, best practices and standards of care exist for management of raptors used as educational ambassadors.[6] Caretakers want the highest quality of life for their captive charges, and the general public also has a deep concern for the care and management of education animals of all species. Best practices in husbandry, display, handling, and management are critical for the health and welfare of the birds used in education. Raptors whose specific needs are addressed are healthier and behave more comfortably in front of audiences (e.g., perching calmly on a gloved hand or perch, foot-tucking, rousing, preening, eating willingly). In turn, this makes the curriculum more effective and positively memorable.

Professional development opportunities are also available for individuals who care for, train, and present raptors in programs to help them maintain the highest level of "best practices" the field offers.[19] Conferences, workshops, and webinars provide opportunities for growth in the areas of captive management, training (i.e., operant conditioning using positive reinforcement), and enhancement of presentation skills.

Environmental education using live raptors not only increases awareness of the presence of raptors in urban landscapes but should also deliver the most up-to-date and scientifically accurate information in an engaging fashion. Educators can encourage audience members of all ages to take action, providing suggestions like the following for how to assist and engage with raptors in the wild:

- taking down soccer nets when not in use
- capping chimneys
- making glass visible to birds
- acquiring information before removing a fledgling
- pointing out raptors to others when on a walk
- using copper or other alternatives to lead ammunition during hunting season

If learners take even the smallest of actions, they become partners and advocates for raptors in an urbanizing world.

Conservation messages can be effectively shared using means other than live raptor programming. Multiple exposures to a topic presented in a variety of ways have a greater chance of being remembered.[20] In fact, pairing other methods with live raptor programs can be an effective strategy to increase the public's environmental literacy. Websites, webinars, and a variety of social media venues can be used to support live programming (figure 16.2 and color plate 13). Using technology to increase learners' exposure to a subject allows them to more thoroughly incorporate it into their understanding.

The Raptor Center has used this approach successfully by creating an online classroom tool ("Raptor Lab") that takes students through the rehabilitation of a sick bald eagle.[21] The students model how a veterinarian would examine the patient, use diagnostics to arrive at a diagnosis, and create a potential treatment plan. Through this process, students discover a key factor that explains why the eagle required rehabilitation, most often due to an anthropogenic cause, which opens up discussions on many levels. The students are exposed to how good science works by being given a question to answer, finding and graphing supporting evidence, and then drawing conclusions. Then they write a research paper and present it to their class.

Following this process, where geographically possible, an educator and live bald eagle visit the classroom, making the lessons learned online relevant and real. Students are also able to look at real data from a rehabilitation clinic and connect it to a live animal. The process can be a very powerful learning experience.

Figure 16.2. Online learning is one method used for raptor conservation programming.

Summary

Urbanization leads to a greater intersection of raptors and human-altered landscapes, and species have to adapt in order to thrive (see chapters 2, 3, 15, and 17). In these areas, the human population, level of public awareness, and incidence of human-raptor interactions continue to grow significantly, resulting in an increased demand for raptor rehabilitation, education, and outreach services. State and federal authorization are required for rehabilitation and professional programming using live raptors, and permits restrict what activities can be lawfully conducted.

A large number of raptors are treated at urban rehabilitation facilities, and a significant amount of information can be collected and analyzed. These data can be used to assist state and federal wildlife agencies in creating mitigation plans for anthropogenic influences on raptors at individual and population levels. Two current barriers to data collection are the lack of consistency within the rehabilitation community and the lack of a central database. The CWHI, created in 2011, is beginning to address these issues.

Conservation-based outreach programs are valuable public resources, and educators use numerous methods to deliver messaging. The most popular method is the use of live raptor ambassadors. Other educational tools—such as websites, webinars, social media venues, and online courses—are employed to increase the number of encounters an individual or a group experiences.

As the fields of raptor rehabilitation and public outreach continue to expand and increase in professionalism, animal welfare standards have risen to the forefront. Best practices, crafted with quality of life and ethical considerations in mind, have led to the development of a variety of university- and college-level training opportunities across the United States.

In an urbanizing world, the importance of raptor rehabilitation, education, and outreach cannot be overstated. Both raptors and humans can be victimized by human-altered landscapes. The aspiration is to provide services to assist both sides: humane care for injured and orphaned raptors; education for care providers; and outreach to answer questions, calm fears, and engage people of all ages to take action to help mitigate the challenges we face together.

Literature Cited

1. International Wildlife Rehabilitation Council. 2010. "About Us." Eugene, OR. Accessed January 21, 2017. https://theiwrc.org/About-us.

2. National Wildlife Rehabilitator's Association. 2008. "History of the NWRA." St. Cloud, MN. Accessed January 21, 2017. http://www.nwrawildlife.org/?page=History.

3. Electronic Code of Federal Regulations. 2017. "Migratory Bird Permits–Rehabilitation Permits." Washington, DC. Accessed January 21, 2017. http://www.ecfr.gov/cgi-bin/text-idx?SID=fa31e11f2b1512a9490feabbe9344b8e&mc=true&node=pt50.9.21&rgn=div5#se50.9.21_131.

4. Deam, S. L., S. P. Terrel, and D. J. Forrester. 1998. "A Retrospective Study of Morbidity and Mortality of Raptors in Florida: 1988–1994." *Journal of Zoo and Wildlife Medicine* 29:160–64.

5. Bystrak, D., E. Nakash, and J. A. Lutmerding. 2012. "Summary of Raptor Banding Records at the Bird Banding Lab." *Journal of Raptor Research* 46:12–16.

6. Arent, L. 2006. *Raptors in Captivity: Guidelines for Care and Management*. Surrey, BC: Hancock House Publishers.

7. Willette, M., J. Ponder, D. L. McRuer, and E. E. Clark, Jr. 2013. "Wildlife Health Monitoring Systems in North America: From Sentinel Species to Public Policy." In *Conservation Medicine: Applied Cases of Ecological Health*, edited by A. Aguirre, 552–62. New York: Oxford University Press.

8. American Veterinary Medical Association. 2017. "One Health." American Veterinary Medical Association, Schaumberg, IL. Accessed January 21, 2017. http://www.avma.org/KB/Resources/Reference/Pages/One-Health94.aspx.

9. Willette, M. 2011. "Clinical Wildlife Health Initiative." The Raptor Center. College of Veterinary Medicine, University of Minnesota, St. Paul, MN. Accessed December 18, 2016. https://www.raptor.umn.edu/our-research/clinical-wildlife-health-initiative.

10. Kelly, T. R., P. H. Bloom, S. G. Torres, Y. Z. Hernandez, R. H. Poppenga, W. M. Boyce, and C. K. Johnson. 2011. "Impact of the California Lead Ammunition Ban on Reducing Lead Exposure in Golden Eagles and Turkey Vultures." *PLoS ONE* 6: e17656. doi:10.1371/journal.pone.0017656.

11. Martell, M., J. Goggin, and P. T. Redig. 2000. "Assessing Rehabilitation Success of Raptors through Band Returns." In *Raptor Biomedicine III*, edited by J. T. Lumeij, J. D. Remple, P. T. Redig, M. Lerz, and J. E. Cooper, 327–34. Lake Worth, FL: Zoological Education Network.

12. American Association of Zoological Veterinarians. 2017. "Community Calendar." Yulee, FL. Accessed January 21, 2017. https://www.aazv.org/events/event_list.asp.

13. American Board of Veterinary Practitioners. 2017. "Annual Symposium." American Board of Veterinary Practitioners, Gainesville, FL. Accessed January 21, 2017. http://abvp.com/symposium.

14. Simmons, B., M. Archie, L. Mann, M. Vymetal-Taylor, A. Berkowitz, T. Bedell, J. Braus, G. Holmes, M, Paden, R. Raze, T. Spence, and B. Weiser. 2004. *Excellence in Environmental Education: Guidelines for Learning (preK–12)*. Washington, DC: North American Association for Environmental Education (NAAEE) Publications.

15. Buhl, G. 2004. *Wildlife in Education: A Guide for the Care and Use of Program Animals*. Edited by G. Buhl and L. Borgia. St. Cloud, MN: National Wildlife Rehabilitator's Association, pp. x, 108.

16. United States Fish and Wildlife Service. 2014. "What You Should Know about a Federal Migratory Bird Special Purpose Possession—Education Permit for Live Birds." USDI Fish and Wildlife Service, Washington, DC. Accessed December 18, 2016. https://www.fws.gov/forms/3-200-10c.pdf.

17. United States Fish and Wildlife Service. 2014. "What You Should Know about a Federal Eagle Exhibition Permit." USDI Fish and Wildlife Service, Washington, DC. Accessed December 18, 2016. https://www.fws.gov/forms/3-200-14.pdf.

18. Carleton-Hug, A., and J. W. Hug. 2010. "Challenges and Opportunities for Evaluating Environmental Education Programs." *Evaluation and Program Planning* 33:159–64.

19. International Association of Avian Trainers and Educators. 2017. "Conferences." International Association of Avian Trainers and Educators, St. Paul, MN. Accessed January 23, 2017. https://www.iaate.org/iaate-conference.

20. Hungerford, H. R., and T. L. Volk. 1990. "Changing Learner Behavior through Environmental Education." *Journal of Environmental Education* 21(3):8–21. doi:10.1080/00958964.1990.10753743.

21. The Raptor Center. 2015. "Raptor Lab." The Raptor Center. College of Veterinary Medicine, University of Minnesota, St. Paul, MN. Accessed January 21, 2017. https://raptorlab.org.

CHAPTER 17

Urban Raptor Case Studies: Lessons from Texas

John M. Davis

THE PRESENCE OF RAPTORS IN cities brings up many questions for wild-life managers and agencies and raises some fundamental questions about how to respond to emerging demands for limited wildlife conservation budgets. According to population estimates, a major demographic shift in the world's human population occurred in May 2007 when we, as a species, became more urban than rural.[1] This rural to urban shift happened in the United States around the late 1910s.[1] In Texas, this shift occurred in the 1940s.[2] As of the time of this writing, the US Census Bureau estimates that more than 80 percent of the US population lives in urban areas. However, the vast urban population of the United States resides on just 3 percent of the land area.[3] This juxtaposition serves as the foundation for one of the most critical issues being debated today in state wildlife agencies regarding the future of wildlife conservation. Where and how should we allocate our limited wildlife conservation resources? Similarly, how are state wildlife agencies to remain relevant to the vast urban public given that 97 percent of the land is rural?

In my experience, the sides of this debate have two differing perspectives. First, the traditional view believes that resources should be focused in rural areas and on game species. This perspective is understandable given that many wildlife managers in state wildlife agencies have come from rural areas, have hunting and fishing backgrounds, and prefer to be in nature away from urban centers. Additionally, the mechanisms that have generated the majority of funding for wildlife conservation for the last 80 years are tied to hunting and angling in rural areas. In 1937, Congress passed the Federal Aid in Wildlife Restoration Act (Pub. L. No. 75–415, 50 Stat. 917 [1937]), which created an excise tax on firearms and ammunition that would be collected at the federal level, then distributed to state wildlife agencies for use in managing and conserving wildlife populations. Since that time, this mechanism has been expanded to include excise taxes on fishing tackle and additional items to fund fish conservation. These acts, frequently referred to as the Pittman-Robertson and the Dingell-Johnson Acts, are primary funding sources for state wildlife and fisheries agencies. However, because the funds are derived primarily from taxes on equipment used in the pursuit of game and fish, they are an incentive for agencies to focus on consumptively used species. Personnel with rural backgrounds and interests in the consumptive use of fish and wildlife and the prevailing funding mechanisms combine to serve as strong forces in the wildlife profession that have created the "game-centric" and rural culture still found in many state wildlife agencies.

However, there is another perspective that recognizes the need for state wildlife agencies to transform their culture to better match that of the citizens of their states.[4] This broader perspective understands that people determine the fate of wildlife and that agencies cannot afford to ignore the wildlife interests and concerns of the vast urban public and hope to remain relevant and viable.[4,5]

As an example of the shift to this perspective, I consider the history of the Texas Parks and Wildlife Department. As an effort to address the wildlife concerns of urban constituents and, therefore, become relevant to urbanites, the department created the Urban Wildlife Program in 1993. The program created offices in the three largest metropolitan areas of the state (Dallas/Ft. Worth, Houston, San Antonio), with each housing two urban wildlife biologists dedicated to wildlife conservation in the assigned metropolitan area. These biologists provided free technical guidance to developers, municipalities, urban landowners and managers, and others to help address all sorts of wildlife and ecological problems encountered in the urban landscape. Very quickly, word spread about the availability of such a valuable resource, and those biologists had more requests for assistance than they could possibly fill. It was clear that urbanites desperately wanted and needed wildlife biologists stationed nearby to help manage urban land and wildlife populations. The program was so successful that it was expanded in 1999: offices were opened in Austin, El Paso, and the Lower Rio Grande Valley.

Over the years, these biologists have learned that many urban wildlife conflicts (or perceived conflicts) can be easily addressed with education and/ or slight behavioral modifications. As a general rule, urban raptors are viewed as beneficial and awe-inspiring wildlife that many citizens are surprised and pleased to encounter in urban areas. As such, addressing urban raptor conflicts usually results in a satisfactory resolution both for the bird and the public. In this chapter, I will describe some examples of situations that urban wildlife biologists have encountered regarding raptors and how they have resolved the issues.

Conflict Resolution Requiring Education Alone

Some urban raptor conflict situations are resolved very easily by educating the citizen. In these cases, the issue is one of simply explaining a bird's behavior or providing a reliable identification and allaying fears. The following examples are representative of these situations.

Example 1: Menacing "Eagle"

One of the Houston urban biologists, Diana Foss, received a call from a frantic homeowner, indicating that there was an eagle perched on her backyard fence. The homeowner had small pets and small children and was afraid to allow any of them outside for fear the eagle would attack them. The homeowner wanted Foss to come and remove the eagle. She did not have the ability to send a photo to Foss, so Foss began the sometimes difficult process of identifying the bird over the phone. Urban biologists quickly learn that one must question callers carefully to get accurate information. Often callers answer questions as they believe they "should" answer rather than by describing what they actually see. Using open-ended questions, Foss guided the homeowner to describe the key features of the "eagle." She asked the homeowner to describe the bird's size in comparison to common items of known size (e.g., softball, football, small child, etc.). She then had the homeowner describe the size of the bill in comparison to known references (e.g., your finger, a ruler, etc.) and asked for general color descriptions as well as descriptions of the legs and feet. At one point, the bird began to walk, so Foss asked the homeowner to describe the walk. Through such open-ended questioning, Foss was able to identify the bird as a yellow-crowned night-heron (*Nyctanassa violacea*). Armed with this information, Foss then educated the homeowner about the bird and allayed any fears of attack upon pets or children. The caller was greatly relieved and no longer wanted the bird removed.

EXAMPLE 2: ACCIPITERS ON THE HUNT

Several urban biologists fielded calls from bird-feeding enthusiasts upset that a Cooper's hawk (*Accipiter cooperii*) or sharp-shinned hawk (*Accipiter striatus*) was taking birds from their feeders. Urban biologists used a two-stage approach to address such calls. First, biologists educated the homeowners about the hawks themselves and encouraged the homeowner to appreciate that the bird-feeding stations were still feeding birds (hawks), just at a higher trophic level than they had perhaps been expecting. Some homeowners absorbed and appreciated this explanation of a more ecologically complex feeding event occurring at their feeders. In these cases, the homeowners' perspectives were broadened, and no action was taken. For cases in which homeowners were still unable to accept hawks' predation of songbirds from their feeders, biologists recommended placing mesh wire or screening around feeders to prevent hawks from having the relatively straight approach lanes needed for successful captures. Surrounding feeding stations with wire mesh with openings large enough to allow songbirds has proven to be a good deterrent.

The presence of Cooper's hawks or sharp-shinned hawks sometimes caused homeowners to notice a sudden disappearance of birds at their feeders. For such reports, urban biologists explained this as a prey response to predator presence and assured them that the songbirds would return when the predator moved to another location.

EXAMPLE 3: AN INJURED RAPTOR?

One of the urban biologists in the Lower Rio Grande Valley, Tony Henehan, received a call from the manager of a retirement community, who indicated that a resident had reported seeing an injured eagle in the birdbath. The manager inquired whether someone could come out and assist. Henehan conducted a site visit; he met with the resident and the manager and learned that the bird was in the birdbath holding its wings outstretched. This had led the resident to believe it was injured, but it had flown away between the time of the call and the site visit. The biologist explained sunbathing behavior to the resident and also asked open-ended questions (as described in example 1) to determine that the bird was not an eagle but a common hawk. The manager and resident were pleased that someone was available to assist them.

Example 4: Mysterious Vulture Behavior

In another situation, a Houston urban biologist, Kelly Norrid, received a call from tenants in a local office building who described a puzzling situation. They noticed periodically that the rubber mats laid out at the front doors of their office complex were rolled up. They did not know the cause until someone noticed black vultures (*Coragyps atratus*) landing and rolling them up. The caller was not upset by this behavior, but rather the office tenants were completely fascinated by it and were trying to learn why the vultures did this. Vultures are known to be attracted to rubber and/or vinyl products.[6] Though no one knows for certain, it has been speculated that chemicals in various rubber products may smell similar to fish or other carrion, though research has not supported this speculation. Norrid passed this information along to the caller, whose curiosity was satisfied. The biologist went on to provide the caller with the details of what could be done to discourage the birds if they ever became a nuisance, though it appeared the birds were not bothersome to the tenants.

Conflict Resolution Requiring Slight Behavioral Modification

Though many conflicts with raptors can be resolved with education alone, some situations require citizens to change their behaviors or do something to accommodate the birds. In many cases, these behavioral changes are minimal and the citizens agree to cooperate once they understand the birds' motivation and the efficacious remedy.

Example 1: Defensive Parents

A San Antonio urban biologist, Richard Heilbrun, received a call from an elderly couple who were unable to enter their backyard without a pair of red-shouldered hawks (*Buteo lineatus*) swooping at their heads in protection of a nest they had constructed in a large oak tree in the yard. Heilbrun conducted a site visit and determined that the hawks had young in the nest, which was the cause of the aggressive actions. He explained the temporal nature of this behavior and the fact that the swooping would likely be over and the young fledged before the homeowners would be able to obtain a permit for nest removal. In the meantime, he advised the homeowners to deter the birds by carrying an umbrella overhead whenever they needed to be in their backyard. He indicated that after the nest was no longer being used by the hawks, the homeowners could remove it. The homeowners were pleased with this solution, accepted the hawks' behavior, and did not pursue removing the nest.

A similar situation occurred at a corporate site in the Dallas/Ft. Worth area. A pair of raptors had built a nest in a tree near the walkway entering the building. Defensive, the birds began diving at people using the walkway. A consulting firm was called, and the biologist identified the problem. Temporary construction fencing was installed to divert foot traffic far enough away from the nest to avoid eliciting the defensive behavior. The redirection of pedestrians away from the vicinity of the nest was successful, and once the young fledged, the fencing was removed.

EXAMPLE 2: RESPONSIBLE ANIMAL OWNERSHIP

In multiple urban areas in Texas, concerned owners of small pets or urban chickens contacted urban biologists about raptors taking these animals. Unfortunately, many of these callers desired permits or other permission to shoot raptors. Biologists informed callers that shooting raptors is not legal and that the problem was an issue of responsible animal ownership. Biologists detailed the various local predators capable of capturing small pets or chickens (raptors, coyotes [*Canis latrans*], raccoons [*Procyon lotor*], foxes [*Vulpes vulpes* or *Urocyon cinereoargenteus*], etc.) and explained that the only way to keep these domestic animals safe would be to construct an enclosure that would protect them on all sides.

EXAMPLE 3: UNWELCOME TENANT

A realty group in the Lower Rio Grande Valley had purchased a property with a warehouse and intended to sell it. While on the property, they noticed that they often flushed a large white owl (barn owl [*Tyto alba*]) in and near a particular part of the building. Concerned about their legal options for removing the owl, they called Henehan. He conducted a site visit to assess the situation and determine if the owl nested in the building. There was no evidence of nesting, and after inspection of a door at one of the warehouse's loading bays, Henehan noticed a hole between the door and the building that allowed the owl entry. He recommended that they patch the hole when the owl was confirmed to be outside the building. They agreed and patched the hole, after which the owl never entered the building again. The realty group was satisfied with the outcome.

Example 4: "Adopted" Vultures

A pair of black vultures nested in flower beds in the courtyard of an office building in San Antonio. The first call received by Heilbrun was an office tenant seeking guidance on what could or should be done. The biologist explained that the birds and nest were federally protected and could not be disturbed. Some of the tenants of the office building not only honored this guidance but moved indoor planters and installed caution tape to block courtyard doors nearest the nest to direct foot traffic to other doors. However, other tenants apparently were not as accepting of the nest and called a pest control company. The second call to Heilbrun was from a tenant upset that a pest control company had been contacted. The biologist conducted a site visit fully expecting to have to reinforce the fact that the birds and nest were protected. However, when he got there, he found that the majority of the tenants loved the birds. With a little discussion and education, the rest of the tenants converted. The offices "adopted" the birds, named them, and developed an office betting pool on the date the eggs would hatch. The nest was successful (figure 17.1), and the vultures returned in subsequent years to successfully fledge young to the excitement of the office tenants.

Example 5: Inadvertently Trapped Raptors

In multiple urban areas across the state, raptors inadvertently found their way into warehouses and ended up perching on rafters. Concerned managers called urban biologists for help. Biologists explained that the most logical way the birds entered the warehouses was by way of large retractable doors. Though it was not certain, the trapped raptors had likely been pursuing prey near one of the large doors, and the momentum of the chase had carried them indoors. Depending on the species, the raptors probably had been pursuing small mammals, sparrows, or insects. Suspecting this, biologists then coached the managers on ways to reverse the process so that the raptors would remove themselves. The specific details varied depending on the facility, but the overall approach was consistent: draw the bird's attention to a nearby retractable door and entice it to fly through to the outdoors. To create prey activity in the opening, workers scattered bird seed in the doorway and immediately outside it. Once the doorway was baited, human traffic through that door was minimized. For a diurnal raptor that hunts birds, the sparrows feeding on birdseed in the doorway often proved irresistible, and the first dive carried the raptor out through the door. Once the bird was out, managers cleaned up the seed to reduce the chance of accidental reentry. If the trapped raptor was nocturnal or typically pursued small mammals, biologists

Figure 17.1. A black vulture nest site in flower beds in the courtyard of an office building in San Antonio, Texas. The adult vulture can be seen behind the lower right corner of the sign, and the half-grown downy nestling is visible at the back of the flower bed in the center of the picture.

recommended waiting until evening to make the conditions most conducive for the bird to leave. In these cases, workers baited the doorway and minimized human traffic as before but also dimmed or turned off the lights in the warehouse and turned on the lights around the nearest retractable door within view of the raptor. The combination of lighting accentuating the door and possible small mammals feeding on the bird seed most often resulted in the raptor flying through that doorway to freedom. If, however, the bird did not exit, a "bait mouse" was sometimes used. Facility staff set up conditions as before and placed a bait mouse ("feeder" mouse purchased from a pet store) in a small, clear container in the doorway. The trapped raptor was attracted to the bait mouse in the doorway, dove at it, and continued out through the door.

Conflict Resolution Requiring Significant Effort

Though most cases of urban raptor conflict can be easily addressed with education or slight modifications, some cases require significant effort to resolve.

EXAMPLE 1: AN IMPROVISED NEST IN A TRASH CAN

A pair of red-shouldered hawks had built a nest in a heavily urbanized city pocket park near the San Antonio Botanical Center close to downtown. The nest had somehow become dislodged and fallen to the ground, but the nestlings had survived the fall. Concerned neighbors contacted the Botanical Center, and city staff took action. They cut a 30 gallon plastic trash can in half horizontally to make it shallow enough to house the nest. They then mounted the trash can about 6 m above the ground on the trunk of the tree from which the nest fell. They placed the nest in the trash can and then moved the nestlings to the nest. The adults returned and not only successfully fledged those young from the artificial nest structure but also fledged young from the same structure for several more years.

EXAMPLE 2: CONSERVATION NETWORKING

About nine in the morning one day, one of the urban biologists in the Dallas/Ft. Worth office, Sam Kieschnick, fielded a call from a truck driver about "a friend" who had hit a bald eagle (*Haliaeetus leucocephalus*) on the road near Cleburne, Texas, which was about 80 km away from the Dallas/Ft. Worth urban office. The truck driver reported that the bird was on the side of the road, limping. Kieschnick phoned a park ranger he knew at Cleburne State Park, which was near the bird's location, and the ranger agreed to drive to the bird, confirm the species, and keep watch until help could arrive. Kieschnick also called the Blackland Prairie Raptor Center (a rehabilitation center) in Allen, Texas, which was located about 160 km from the bird's location. It was quickly determined that the best facility in the state with the expertise to handle such a powerful bird was Last Chance Forever in San Antonio, run by master falconer John Karger. Though that facility was more than 355 km away, biologists scrambled a team to get this bird. All parties converged at the eagle's location. The team secured the bird, and the capture was photographed and posted on agency social media sites. The posts went viral, local news channels picked up the story, and biologists disseminated a great deal of educational information. The eagle was taken to Last Chance Forever, where all efforts to save it unfortunately failed. Though the outcome was not what everyone had hoped, the network of conservationists worked very well together to attempt to save this bird. Having such a well-functioning network in place for events like this is often critical for the treatment and release of injured urban raptors.

Example 3: Unwanted Vultures Getting the Message

In the Houston area, Norrid regularly received calls about vultures roosting in large numbers on communication towers. Callers were typically concerned about the accumulation of excreta at the base of towers. Some research related to this phenomenon[7,8,9] has indicated that effigies (crude models of vultures) hung upside down on towers can be effective at greatly reducing or eliminating vultures at a particular site. One effigy per tower, hung as low as 30 m or less, can be sufficient.[9] Norrid recommended this practice and provided callers with a supplier of effigies for movies. After years of recommending this practice and encouraging callers to contact him for further assistance as needed, he has not received a single follow-up call.

Example 4: Nest Tree Removal

Finally, it is not uncommon to field calls in all of these urban areas from landowners or project managers who need (or want) to take down a tree housing an active raptor nest (i.e., one containing eggs or young). In these cases, the biologists educate the callers regarding the protection of the birds and the nest and advise them to halt the project until the nest is no longer active, when it can then be removed. If this option is not satisfactory, the callers are directed to the US Fish and Wildlife Service to pursue the appropriate permits. The option chosen varies based on the flexibility of the construction schedule and other factors peculiar to each project.

The Growing Need for Urban Wildlife Expertise

These case studies highlight the growing need for urban wildlife biologists operating in cities across the nation to help citizens address urban wildlife issues. The Texas Urban Wildlife Program experience has proven the demand for such expertise. The program is very popular, and urban biologists have had to become skillful at selecting projects to maximize impact while minimizing time commitment. It was this need to maximize efficiency that led the Urban Wildlife Program and the city of San Antonio to create the Texas Master Naturalist Program in March 1997. By 1998, the Texas Master Naturalist Program had grown to a statewide initiative sponsored by the Texas Parks and Wildlife Department and Texas A&M AgriLife Extension Service. The core mission of this program is to develop a corps of well-informed volunteers to provide education, outreach, and service dedicated to the beneficial management of natural resources and natural areas within their communities for the state of Texas. By partnering with many conservation organizations to develop this corps, urban wildlife biologists were able to multiply conservation efforts in urban areas significantly. At the time of this writing, the program has 48 chapters across the state of Texas and 28 states have developed similar programs modeled on the Texas Master Naturalist Program.

The Texas Urban Wildlife Program has also become a national leader among state wildlife agencies, and our biologists have begun to mentor those in other states developing similar programs. As state agencies continue to recognize the need to manage urban wildlife, mitigate human-wildlife conflict, and be relevant among their urban constituents, programs like these provide sought-after services demonstrating the value of the state wildlife agency to the urban public.

Acknowledgments

I thank urban wildlife biologists Diana Foss, Tony Henehan, Kelly Norrid, Richard Heilbrun, and Sam Keischnick for providing details of the accounts from their respective cities covered in this chapter. I also thank Michelle Haggerty for providing information regarding the Texas Master Naturalist Program. Finally, I thank the editors Clint Boal and Cheryl Dykstra for valuable assistance in developing this chapter.

Literature Cited

1. North Carolina State University. 2007. "Mayday 23: World Population Becomes More Urban than Rural." *ScienceDaily*. Accessed December 12, 2016. http://www .sciencedaily.com/releases/2007/05/070525000642.htm.

2. United States Census Bureau. 1995. "Urban and Rural Population: 1900 to 1990." US Census Bureau. Accessed February 14, 2017. http://www.census.gov/population/ censusdata/urpop0090.txt.

3. United States Census Bureau. 2015. "New Census Data Show Differences between Urban and Rural Populations." US Census Bureau American Community Survey. Accessed December 12, 2016. http://www.census.gov/newsroom/press-releases/ 2016/cb16-210.html.

4. Association of Fish and Wildlife Agencies. 2016. "The Future of America's Fish and Wildlife: A 21st Century Vision for Investing in and Connecting People to Nature." Association of Fish and Wildlife Agencies. Accessed March 24, 2017. http://www .fishwildlife.org/files/Blue_Ribbon_Panel_Report2.pdf.

5. Adams, C. E., K. J. Lindsey, and S. J. Ash. 2006. *Urban Wildlife Management.* Boca Raton, FL: CRC Press.

6. USDA. 2016. "Managing Vulture Damage." US Department of Agriculture Wildlife Services. Accessed February 15, 2017.https://www.aphis.usda.gov/publications/ wildlife_damage/content/printable_version/fs_vulture_damage_man.pdf.

7. Avery, M. L., J. S. Humphrey, E. A. Tillman, K. O. Phares, and J. E. Hatcher. 2002. "Dispersing Vulture Roosts on Communication Towers." *Journal of Raptor Research* 36:45–50.

8. Seamans, T. W. 2004. "Response of Roosting Turkey Vultures to a Vulture Effigy." *Ohio Journal of Science* 104:136–38.

9. Ball, S. A. 2009. "From the Field: Suspending Vulture Effigies from Roosts to Reduce Bird Strikes." *Human-Wildlife Conflicts* 3:257–59.

CHAPTER 18

Management and Conservation
of Urban Raptors

David M. Bird, Robert N. Rosenfield, Greg Septon, Marcel

A. Gahbauer, John H. Barclay, and Jeffrey L. Lincer

A S MANY CHAPTERS IN THIS book have discussed, whether increased urbanization has overall benefits for raptors is not yet clear. In fact, the fundamental ecology of urban raptors is poorly understood, especially on a long-term basis.[1,2,3]

The urban environment includes hazards that may be more common than in rural settings, many of which were covered in chapters 14 and 15. The density of structures and dynamics of activities in cities likely heighten the chances of collisions between birds and vehicles, aircraft, buildings, powerlines, or other structures. Further, fledgling raptors landing in dangerous environments, electrocution, persistent organic pollutants and heavy metals, and pigeon (*Columba livia*) and rodent poisoning programs all pose viable threats. Even bridge and building owners and managers display mixed feelings toward having peregrine falcons (*Falco peregrinus*) "adopt" their structures for nesting. Messy food remains, aggression toward building staff (e.g., window cleaners) leading to work delays, and the unwanted responsibility of catering to a species of concern for the government can lead to animosity toward the birds. Indeed, not all members of the general public welcome raptors into their towns. Claims of hawks and falcons affecting numbers of songbirds visiting feeders and concerns about raptors defending their nests by striking humans or preying on small pets are frequently heard by city officials.[4,5]

However, as presented earlier in this book, raptors may also benefit from settling in urban environments. The abundance of almost year-round prey such as feeder birds, pigeons, and squirrels (Sciuridae) provides stability of food resources, which potentially reduces or even eliminates the need for migration, a risky and energetically expensive process. This can lead to earlier nesting, and thus a longer period for juveniles to acquire flight and hunting skills or an enhanced opportunity for renesting after a failed attempt. For species that can adapt to nesting on buildings and other structures, cities offer an abundance of well-sheltered nesting sites often inaccessible to mammalian predators. Large cities may also offer thermal advantages to individual birds, because cities are generally warmer than their surroundings.[6] However, the vegetation communities in urban settings may influence species differently. In general, the probability of raptor presence in urban areas was negatively associated with the extent of lawn cover.[7] But when looking at individual species, occupancy probabilities of red-tailed hawks (*Buteo jamaicensis*) and Cooper's hawks (*Accipiter cooperii*) were positively associated with amount of woodland cover, and American kestrel (*F. sparverius*) occupancy was positively associated with extent of grassland cover. It is evident, and not surprising, that different raptor species will perceive and use urban areas differently.

In an attempt to better facilitate conservation of raptors in urban environments, this chapter reviews certain aspects of the current knowledge of urban raptor ecology and provides some specific management recommendations for biologists and city managers. Accordingly, we present case histories regarding conservation of three raptors—the peregrine falcon, the Cooper's hawk, and the burrowing owl (*Athene cunicularia*). We chose these species because they represent diverse challenges for urban wildlife management. We suggest that the relatively common occurrence of these raptors in cities makes them excellent ambassadors for public education on urban wildlife. Raptors also provide ready opportunities for much-needed research on their urban ecology. Additional knowledge will both enhance our understanding of raptors' urban roles and suggest appropriate conservation measures.

Case History One: Managing the Peregrine Falcon in an Urban Core Environment

Although peregrine falcons are known to have nested on human-built structures in Europe as far back as the Middle Ages,[8] this behavior was first documented in North America in 1936. From that year through 1952, a pair nested annually on the Sun Life Building in Montreal, Quebec. Eggs rolled off the building each spring until 1940, when a shallow nest tray filled with sand and gravel was provided to increase the chance of successful nesting,[9] thus constituting the very first recorded act of managing raptors in an urban setting!

The colonization of urban areas by raptors began with the release of captive-bred peregrines in cities in the 1970s.[10,11] Individuals released in cities preferentially returned to nest in similar environments, and presumably their offspring experienced a similar inclination. The scarcity of natural predators such as great horned owls (*Bubo virginianus*) also may have enhanced the breeding success of city falcons. The urban population of peregrines has grown steadily over the past four decades to the point where several major cities including Chicago, Toronto, Montreal, and New York now each host a dozen or more nesting pairs annually, and smaller numbers exist in other North American cities such as Omaha, Saskatoon, and Winnipeg. Some of these cities are hundreds of kilometers from any historical cliff-nesting locations. In Wisconsin today, for example, more than 80 percent of peregrine nests are located on human-built structures in urban settings.

Besides the absence of nest predators, the abundance of prey and nest shelter and substrate also makes cities attractive to peregrines. The diversity of species hunted by urban peregrines is surprisingly large; one review tallied 104 bird species taken as prey at 19 urban nests.[8] More significantly, urban prey populations are typically robust throughout the year, whereas in natural environments there can be considerable seasonality to prey, and even during the breeding season, it may be scarcer than in cities.

As foreshadowed by the Sun Life Building peregrine situation, human intervention in making nest ledges more suitable has also proven to be a significant factor in the success of urban peregrines. In a study of 87 urban nest sites in eastern North America, the total number of young fledged was almost three times higher at nest sites with gravel-lined trays or boxes than at those lacking them.[12] The gravel provides a secure substrate for eggs, and boxes can shelter incubating adults, eggs, and young from inclement weather.

Of course, cities also pose certain risks to peregrines that are scarce or absent in natural areas.[13] A review of 160 reported peregrine mortalities in Ontario,

Massachusetts, and Pennsylvania from 1988 to 2006 found that nearly 40 percent were a result of collisions with buildings.[12] This hazard is largely a result of many downtown buildings being built of reflective glass[14] and can be exacerbated by wind shear created through the artificial urban canyons formed by skyscrapers. Fledging from buildings is frequently a risky occasion, as the paucity of perches other than the nest ledge often leads many young birds to "helicopter" down to street level where they are exposed to traffic, attacks from terrestrial predators, and potential starvation because their parents are reluctant to come to the ground to feed their offspring. Fortunately, many urban peregrine nests are watched over by dedicated local volunteers who step in to perform rescues in such situations. For example, in southern Ontario, 44 percent of young fledged between 1995 and 2006 were rescued at least once.[15]

Urban peregrines face various other risks. Young may become trapped in chimney vents or other tight spaces from which they are unable to escape without assistance. Transmission lines, vehicles, and aircraft in city airports can pose a collision risk for peregrines pursuing prey in an otherwise open right-of-way. Some adults have been killed through secondary poisoning, usually from feeding on feral pigeons who have eaten poisoned bait. There have also been occasional reports of peregrines persecuted by pigeon-fanciers angry at the falcons killing their pigeons (e.g., the Watts District of Los Angeles).[11] Despite these various hazards, the steadily increasing urban population of peregrines over the past four decades suggests that urban areas, overall, are productive environments for this species.[11,16]

However, nesting on bridges can often be disastrous for peregrines. In several cities including Montreal, New York, and Philadelphia, peregrines have taken to nesting under many of the large highway bridges. Unfortunately, fledging success tends to be much lower from these nests. Some of the young find themselves in the water on their first flights and drown, whereas others perch on the bridge railings and end up as road kill. Mean annual productivity at bridge nests was lower by approximately 0.5 young per nest than at other urban locations.[12] Placing nesting boxes on the bridge supports that are over land can encourage the peregrines to nest in a somewhat safer location.

Much of the management of urban peregrines relates to providing suitable shelter and substrates to enhance probability of nesting success and intervening to rescue (and if necessary, rehabilitate) juveniles that have come to the ground on one of their early flights. However, it is important to acknowledge that peregrine-human conflicts can also occur on occasion, typically when people have access to ledges or rooftops near where peregrines are nesting.

The degree of territoriality varies among peregrines, but the most aggressive individuals will not hesitate to strike intruders. Conflict can often be limited by having building managers restrict human access during the nesting season, as well as ensuring that any personnel who do need to enter the territory for essential work are trained in how to assess peregrine behavior and equipped with appropriate personal protection equipment. In rare cases where regular access to the nest is required or a site is deemed unsafe for the birds, it may be necessary to actually move a nest by relocating an existing nest structure to another suitable location on a nearby building. Researchers (the first author of this chapter and collaborator I. Ritchie) successfully relocated one Montreal nest to another building three blocks away by removing the live fertile eggs from an undesirable nesting site (which was then rendered inaccessible) and placing them in an incubator. Meanwhile, a dummy clutch composed of infertile peregrine eggs mixed with brown chicken eggs was placed in a nest tray located on another building ledge within obvious view of the female. Within just hours, the female adopted the "dummy clutch," which was later switched with her original "live clutch." Other management options include installing a new shelter to enhance the attractiveness of an alternate location or reducing the suitability of the existing site (e.g., blocking access). Building managers who are sympathetic to peregrines can also help by installing and maintaining nest boxes, as well as providing security.

The growing presence of peregrines in cities has provided great opportunities for public education. Dozens of urban nests have been featured on popular webcams with followers all over the world. Many locations have also attracted visitors to ground-level information centers featuring live viewing of the nests on large screens and public events to showcase banding of the offspring, which provide opportunities to give information on the natural history of peregrines and explain the value of ongoing monitoring efforts. The conservation success of this species combined with the story of its adaptation to urban environments has formed the foundation of countless classroom presentations that tie into curriculum topics ranging from food webs and bioaccumulation to the geography of migration and the physics of the peregrine's high-speed hunting dives. As the number of nest sites has grown over time in many cities, major media outlets still provide ongoing coverage and public awareness of the species as it continues to expand. Chicago declared the peregrine the city's official bird in 1999, and in many other eastern North American cities, it is now also a well-recognized and much-loved resident. Through their high profile, peregrines have also shed light on the broader phenomenon of urban raptors.

Case History Two: Managing the Cooper's Hawk in a Backyard Suburban Environment

The Cooper's hawk, a crow-sized raptor that typically preys on small to medium-sized songbirds and doves, may be the most common urban-breeding raptor throughout much of North America since around the 1980s.[17] Indeed, as shown in chapter 7, this raptor exhibits marked flexibility in urban breeding habitat, which ranges from concrete-dominated landscapes to suburban developments.[18,19] Unlike other raptor species that can be inadvertently "managed" in part by the presence and placement of human-made structures used for nests,[1] nesting Cooper's hawks apparently are averse to using anything other than trees.[20]

Cooper's hawks are present year-round in many cities throughout the contiguous United States and in some cities in southern Canada.[21] Although some human-related mortality factors such as shooting, electrocution, and collisions with buildings and vehicles may be more frequent in urban settings,[22] Rosenfield and his coworkers found that annual survivorship was relatively high (approximately 80 percent) and did not differ between breeding male Cooper's hawks in urban versus rural Wisconsin.[23] Moreover, Cooper's hawks nesting in high densities in cities like Victoria, British Columbia,[24] and metropolitan Milwaukee, Wisconsin,[20] may be thriving on the abundance of prey in cities, which is often higher than in exurban environments.[25] Telemetry studies generally reveal that urban breeding and wintering Cooper's hawks do not have to range as far as rural individuals to obtain adequate food.[26] Although introduced species, such as house sparrows (*Passer domesticus*), European starlings (*Sturnus vulgaris*), and domestic rabbits (*Oryctolagus cuniculus*) cause damage to native ecosystems and species, they can also constitute important prey for Cooper's hawks.[24]

The importance of bird feeders to the diet of urban Cooper's hawks remains unclear. In a review of Project Feederwatch data from across North America, 25 species of predators were reported at feeders, with Cooper's hawks being among the top three.[27] The apparent importance of feeders to Cooper's hawks may be biased, though, because people may see them most often at bird feeders. For example, there was no systematic use of bird feeders by Cooper's hawks wintering in Terra Haute, Indiana, according to Roth and coworkers.[21] Moreover, they suggested that the lack of a strong tendency to hunt around feeders may constitute a form of "prey management" by Cooper's hawks, whereby they avoid repeated attacks at areas (feeders) frequently visited by behaviorally responsive prey, thus facilitating a source of prey over the long term.

The practice of bird feeding affects all aspects of avian biology, particularly with respect to disease transmission (e.g., salmonellosis, aspergillosis, trichomoniasis).[28] Although health of prey has the potential to impair reproductive success or heighten the risk of mortality for urban Cooper's hawks, it is not consistent from city to city. High levels of trichomoniasis were recorded among urban Cooper's hawk nestlings in Tucson, Arizona, likely due to high consumption of abundant columbid (pigeons and doves) prey.[18] However, other studies in Victoria, British Columbia; Grand Forks, North Dakota; and Stevens Point, Wisconsin, reported no deaths of Cooper's hawks due to trichomoniasis.[29] In the United States, urban birds, including feeder birds, may be the most prevalent carrier species for West Nile virus (WNV),[30,31] and these birds may be preferred for feeding by mosquito (*Culex* spp.) vectors. In one study of urban raptors in metropolitan Milwaukee, Wisconsin, Stout and coworkers found WNV antibodies in 88 percent of breeding adult and 2.1 percent of nestling Cooper's hawks, but with no detectable adverse effects on that population.[32] Cooper's hawks admitted to a rehabilitation center in Illinois had antibody levels against the bacterium *Mycoplasma gallisepticum*, a pathogen carried by feeder birds that causes conjunctivitis. Their levels were twice as high as several other raptors admitted to the center, but there were no physical signs of infection in any raptor sampled.[33] We note, however, that few bird-feeder studies exist and it is still unclear how feeders might affect avian infection risk dynamics in cites.[34]

Urban areas can also present indirect threats to raptors via chemical pollution from heavy metals, rodenticides, and persistent organic pollutants.[35] Samples of Cooper's hawks in and around Vancouver, British Columbia, from 1999 to 2010 exceeded the highest concentrations of polybrominated diphenyl ethers (PBDEs), commonly known as flame retardants, reported in the literature for wild birds, with higher concentrations in urban birds than rural ones.[35]

Finally, Cooper's hawks can be aggressive near their nests, and infrequently, an individual bird or pair will dive-bomb (and sometimes strike) people who venture near nest trees.[18,36] Fortunately, this appears to be a rare occurrence and usually only occurs for the two to four week period just before the young leave the nest. City employees, avian researchers, and wildlife officials typically have been successful at markedly reducing this risk by rerouting human traffic away from nest trees during the nestling stage.[5] In public places, it is important to erect placards on-site to explain the phenomenon; we also recommend personal meetings between landowners (private or otherwise) and wildlife officials who can explain the risk and the biology of the hawks as well as potential mitigation.

Case History Three: Managing the Burrowing Owl in Open Suburban Landscapes

Burrowing owls breed in climax shortgrass prairies, including a variety of disturbed grasslands and short mixed-herbaceous plant communities where various agents of disturbance maintain the short, open-vegetation conditions the owls require. Besides grazing by large and small mammals, land and vegetation management practices such as mowing, shallow disking (i.e., shallow enough to protect underground burrow systems), herbicide application, and fire also function to maintain disclimax communities by removing standing vegetation and maintaining the short, sparse vegetation conditions that burrowing owls select as nesting areas.[37]

As also shown in chapter 12, the process of conversion of rural and agricultural lands to suburban and urban land uses often unintentionally creates owl habitat where it did not previously exist. Typically, the first phase of developing agricultural lands is the cessation of tilling or other crop management practices and removing trees where orchards existed. Until development occurs, the lands usually remain idle except for activities to remove and/or manage the ruderal vegetation that develops in the absence of tilling. With the commercial development of agricultural lands in California, ground squirrels (*Otospermophilus beecheyi*) usually colonize such abandoned lands, thereby providing burrows for owls. As a result, burrowing owl colonies become established, and the process often leads to the unintended result of ephemeral or temporary owl colonies. As another example, on a large geographic scale, the clearing of woodland and draining of wetlands in Florida created habitat that enabled the owl's population to increase and its range to expand.[38]

The creation of ephemeral burrowing owl colonies has been further aggravated by ordinances in several California municipalities that require landowners to reduce fire hazards on vacant lands. Disking is a common practice to remove standing herbaceous vegetation, but disking destroys burrow entrances and can entomb owls in burrows. Disking to reduce fire hazards is typically done in the early stages of the dry season (e.g., about July 1 in California), which is when nesting owls have young. Thus, mortality associated with disking can be substantial. Conservationists persuaded some municipalities in California (e.g., San Jose, Davis) to adopt ordinances prohibiting disking as a means of fire hazard reduction on vacant lands and, thereby, reducing direct mortality of owls, eggs, and/or nestlings.

As a more expensive alternative to disking, mowing removes standing vegetation while not destroying burrows. The regular mowing of airport infields, nonirrigated margins of golf courses, and suburban or urban vacant lands unintentionally maintains short, sparse vegetation and ideal conditions for nesting burrowing owls. Consequently, burrowing owls often occur on airports within their range in the western United States and Florida[39,40] and on golf courses, fairgrounds, cemeteries, and highway cloverleaf intersections in suburban and urban settings.[37,40]

The phenomenon of unintentionally creating and maintaining owl nesting habitat in the urban/suburban/rural matrix has contributed to temporary local population increases. However, these increases are often ephemeral in nature and are followed by declines. This is because the activities that created and maintained environments suitable for owls are relatively short-term practices associated with initial stages of conversion of agricultural land to development.

Millsap and Bear[41] studied the nest density and reproduction of burrowing owls along a gradient where development ranged from less than 2 percent to more than 80 percent of the landscape in Florida. They found that owl nest-site density and fecundity (fledged young per nest site) increased until 45–60 percent development occurred, beyond which both metrics declined. Nest success did not vary along the development gradient, but the proportion of nests that failed due to human activities increased with escalating development. To manage and conserve owls on vacant lands, the researchers recommended no-disturbance buffer zones, even as small as 10 m, around nests to allow owls to nest successfully and fledge more young. However, buffer zones can have unwanted consequences if they exclude whatever agents (livestock or mowers) maintained the short, sparse vegetation that initially attracted the owls. Consequently, buffer zones should be used with great caution and employed only under the direct supervision of a raptor biologist experienced in monitoring burrowing owl nesting behavior.

The occurrence of owls on lands intended for development presents a management challenge because something must be done about the owls prior to the final conversion that makes the land unsuitable to them. The most widespread practice in California has been to passively relocate owls from occupied land prior to its final conversion, thereby avoiding direct mortality.[42] Passive relocation involves forcing owls to vacate occupied sites by blocking burrows with one-way devices, thus allowing owls to exit burrows but not reenter them. This technique avoids the legal issues of capturing and transporting the birds, practices that are prohibited by various regulations and policies. Passive relocation can be effective on small scales, such as along utility corridors, to minimize owl occurrence inside burrows and avoid direct mortality when such lands are finally graded. We are, however, not aware of any evidence that passive relocation has been widely successful at establishing relocated owls on new sites, based on observable results, such as owls nesting there in successive years.

The passive relocation technique also does not address the regulatory requirement to mitigate for the loss of owl habitat as required by the National Environmental Policy Act (NEPA) and the California Environmental Quality Act (CEQA). A common practice in California has been to set aside burrowing owl habitat on preserves, reserves, or "mitigation sites" to compensate for the loss of habitat on development sites. Although this practice has fulfilled regulatory requirements, in reality it has conspicuously failed to provide environments that owls actually use for nesting or to address the regional declines in owl populations in California.[43,44] This is largely due to the selection of mitigation sites in less-expensive areas outside the region where habitat was lost. An unpublished survey of mitigation sites to compensate for the loss of owl habitat in the Santa Clara Valley by the Santa Clara Valley Audubon Society failed to find evidence that mitigation sites supported nesting owl populations comparable to those nesting in the habitat that was created by human activities but subsequently lost, according to researcher C. Breon. Another shortcoming of the mitigation site practice is the failure to manage the vegetation on mitigation sites so as to provide appropriate nesting conditions for the owls. There needs to be recognition that some form of disturbance (e.g., grazing, mowing, fire, disking) is necessary to maintain the disclimax conditions that owls use for nesting.

Burrowing owls have rather modest habitat requirements, and they are tolerant of human activity around their nesting sites. This would suggest managing and maintaining owl colonies in urban and suburban settings should be straightforward. However, to maintain owl colonies in urban settings requires an understanding of the ecological processes that lead to, and maintain, nesting and foraging habitat and implementation of regulatory mechanisms to continue appropriate ongoing disturbance regimes.

Summary and Recommendations

A wide variety of raptors have adapted to urban landscapes, and their continued success can be supported through adoption of effective management and conservation practices. Active involvement with nesting raptors can include the following:

- providing well-placed nest boxes or platforms to improve nesting success; in the case of bridges, they should be situated over land to reduce the risk of fledglings drowning;
- fostering and maintaining positive working relationships with building owners and managers so that they also derive benefits through public relations and education;
- protecting active nest sites from human disturbance and concurrently reducing the risk of raptor aggression toward people by establishing buffer zones around them while making an effort to inform local residents or tenants about the ecology and behavior of the birds;
- training personnel who need to perform essential work within active raptor nesting territories so that they can assess the birds' behavior, and equipping them with appropriate personal protection equipment;
- moving raptor nests from undesirable locations through means such as installing a new nest box or platform to enhance the attractiveness of an alternate location, reducing the suitability of the existing site (e.g., blocking access), or even translocating nests; and
- having experienced biologists carry out management activities such as passive relocation, exclusion from nests or burrows, erection of exclusion zones, or development of site- and species-specific management plans.

More generally, the urban environment can be made safer and more attractive for raptors through taking these steps:

- preserving natural environments for the benefit of raptors and their prey;
- encouraging architects, businesses, and construction companies to reduce raptor collisions through safer design, such as a reduction of mirrored glass on new buildings, installation of bird-friendly glass, or placement of commercially available bird tape on existing glass; and
- working with political authorities, local conservation groups, and media to campaign against pigeon-poisoning programs and other chemical pollution, including the use of rodenticides, persistent organic pollutants, and heavy metals.

The growing presence of urban raptors also offers great potential for further research and education. For example, the effect of bird feeders on urban raptors remains unclear, and further research is needed to better understand their importance as prey bases and centers of disease transmission. Additionally, wildlife rescue centers provide not only a valuable service for helping injured urban raptors and facilitating public awareness of them, but also opportunities to quantitatively assess the relative threats to these birds. Finally, the proliferation of webcams, information centers, classroom education programs, and media coverage of urban raptors provides unparalleled opportunities for public education at all age levels on a variety of conservation themes (e.g., persecution, habitat preservation, environmental health, climate change). The relatively new field of "urban ornithology" is rapidly expanding,[45] and raptors provide an excellent opportunity to better understand ecological processes within this environment, including human activities that can enhance or harm raptor populations.

Literature Cited

1. Bird, D. M., D. Varland, and J. J. Negro, eds. 1996. *Raptors in Human Landscapes: Adaptations to Built and Cultivated Environments*. San Diego: Academic Press.
2. Love, O. P., and D. M. Bird. 2000. "Raptors in Urban Landscapes: A Review and Future Concerns." In *Raptors at Risk: Proceedings of the V World Conference on Birds of Prey and Owls*, edited by R. D. Chancellor and B.-U. Meyburg, 425–34. Berlin, Germany: World Working Group on Birds of Prey and Owls; Surrey, BC: Hancock House Publishers.
3. Rutz, C. 2008. "The Establishment of an Urban Bird Population." *Journal of Animal Ecology* 77:1008–19.
4. See chapter 15.
5. See chapter 17.
6. Landsberg, H. E. 1981. *The Urban Climate*. New York: Academic Press.
7. Hogg, J. R., and C. H. Nilon. 2015. "Habitat Associations of Birds of Prey in Urban Business Parks." *Urban Ecosystems* 18:267–84.
8. Cade, T. J., M. Martell, P. Redig, G. A. Septon, and H. Tordoff. 1996. "Peregrine Falcons in Urban North America." In *Raptors in Human Landscapes: Adaptations to Built and Cultivated Environments*, edited by D. Bird, D. Varland, and J. J. Negro, 3–13. San Diego: Academic Press.
9. Ratcliffe, D. 1980. *The Peregrine Falcon*. Vermillion: Buteo Books.
10. Holroyd, G. and D. M. Bird. 2012. "Lessons learned during the recovery of the Peregrine Falcon in Canada." *Canadian Wildlife Biology and Management* 1:3–18.
11. See chapter 13.
12. Gahbauer, M. A., D. M. Bird, K. E. Clark, T. French, D. W. Brauning, and F. A. McMorris. 2015. "Productivity, Mortality, and Management of Urban Peregrine Falcons in Northeastern North America." *Journal of Wildlife Management* 79:10–19.

13. Cade, T. J., and D. M. Bird. 1990. "Peregrine Falcons, *Falco peregrinus*, Nesting in an Urban Environment: A Review." *Canadian Field-Naturalist* 104:209–18.

14. Klem, D., Jr. 1989. "Bird-Window Collisions." *Wilson Bulletin* 101:606–20.

15. Gahbauer, M. A., D. M. Bird, and T. R. Armstrong. 2015. "Origin, Growth and Composition of the Recovering Peregrine Falcon Population in Ontario." *Journal of Raptor Research* 49:281–93.

16. Septon, G. A., J. Bielefeldt, T. Ellestad, J. B. Marks, and R. N. Rosenfield. 1996. "Peregrines Falcons, Power Plant Nest Structures and Shoreline Movements." In *Raptors in Human Landscapes: Adaptations to Built and Cultivated Environments*, edited by D. Bird, D. Varland, and J. J. Negro, 145–54. San Diego: Academic Press.

17. Curtis, O. E., R. N. Rosenfield, and J. Bielefeldt. 2006. "Cooper's Hawk (*Accipiter cooperii*)." In *The Birds of North America*, edited by P. G. Rodewald. Ithaca: Cornell Lab of Ornithology. Accessed March 9, 2017. https://birdsna.org/Species-Account/bna/species/coohaw.

18. Boal, C. W., and R. W. Mannan. 1998. "Nest-Site Selection by Cooper's Hawks in an Urban Environment." *Journal of Wildlife Management* 62:864–71.

19. See chapter 7.

20. Stout, W. E., and R. N. Rosenfield. 2010. "Colonization, Growth, and Density of a Pioneer Cooper's Hawk Population in a Large Metropolitan Environment." *Journal of Raptor Research* 44:255–67.

21. Roth, T. C., II, W. E. Vetter, and S. L. Lima. 2008. "Spatial Ecology of Wintering *Accipiter* Hawks: Home Range, Habitat Use, and the Influence of Bird Feeders." *Condor* 110:260–68.

22. Chiang, S. N., P. H. Bloom, A. M. Bartuszevige, and S. E. Thomas. 2012. "Home Range and Habitat Use of Cooper's Hawks in Urban and Natural Areas." In *Urban Bird Ecology and Conservation*, edited by C. A. Lepczyk and P. S. Warren. Studies in Avian Biology, no. 45. Berkeley: University of California Press, 1–16. http://www.ucpress.edu/go/sab.

23. Rosenfield, R. N., J. Bielefeldt, L. J. Rosenfield, T. L. Booms, and M. A. Bozek. 2009. "Survival Rates and Lifetime Reproduction of Breeding Male Cooper's Hawks in Wisconsin, 1980–2005." *Wilson Journal of Ornithology* 121:610–17.

24. Cava, J. A., A. C. Stewart, and R. N. Rosenfield. 2012. "Introduced Species Dominate the Diet of Breeding Urban Cooper's Hawks in British Columbia." *Wilson Journal of Ornithology* 124:775–82.

25. Marzluff, J. M., F. R. Gehlbach, and D. A. Manuwal. 1998. "Urban Environments: Influences on Avifauna and Challenges for the Avian Conservationist." In *Avian Conservation: Research and Management*, edited by J. M. Marzluff and R. Salabanks, 283–99. Washington, DC: Island Press.

26. Millsap, B. A., T. F. Breen, and L. M. Phillips. 2013. "Ecology of the Cooper's Hawk in North Florida." *North American Fauna* 78:1–58.

27. Dunn, E. H., and D. L. Tessaglia. 1994. "Predation of Birds at Feeders in Winter." *Journal of Field Ornithology* 65:8–16.

28. Robb, G. N., R. A. McDonald, R. A. Chamberlain, and S. Bearhop. 2008. "Food for Thought: Supplementary Feeding as a Driver of Ecological Change in Avian Populations." *Frontiers in Ecology and the Environment* 6:476–84.

29. Rosenfield, R. N., J. Bielefeldt, L. J. Rosenfield, S. J. Taft, R. K. Murphy, and A. C. Stewart. 2002. "Prevalence of *Trichomonas gallinae* in Nestling Cooper's Hawks among Three North American Populations." *Wilson Bulletin* 114:145–47.

30. Kilpatrick, A. M., L. D. Kramer, M. J. Jones, P. P. Marra, and P. Daszak. 2006. "West Nile Virus Epidemics Are Driven by Shifts in Mosquito Feeding Behavior." *PLoS Biol* 4(4): e82. doi:10.1371/journal.pbio.0040082.

31. Kilpatrick, A. M. 2011. "Globalization, Land Use, and the Invasion of West Nile Virus." *Science* 334:323–27.

32. Stout, W. E., A. G. Cassini, J. K. Meece, J. M. Papp, R. N. Rosenfield, and K. D. Reed. 2005. "Serologic Evidence of West Nile Virus Infection in Three Wild Raptor Populations." *Avian Diseases* 49:371–75.

33. Wrobel, R. R., T. E. Wilcoxen, J. T. Nuzzo, and J. Seitz. 2016. "Seroprevalence of Avian Pox *Mycoplasma gallisepticum* in Raptors in Central Illinois." *Journal of Raptor Research* 50:289–94.

34. Martin, L. B., and M. Boruta. 2014. "The Impacts of Urbanization on Avian Disease Transmission and Emergence." In *Avian Urban Ecology, Behavioral and Physiological Adaptations*, edited by D. Gil and H. Brumm, 116–28. Oxford, UK: Oxford University Press.

35. Elliott, J. E., J. Brogan, S. L. Lee, K. G. Drouillard, and K. H. Elliott. 2015. "PBDEs and Other POPs in Urban Birds of Prey Partly Explained by Trophic Level and Carbon Source." *Science of the Total Environment* 524:157–65.

36. Cartron, J.-L. E., P. L. Kennedy, R. Yaksich, and S. H. Stoleson. 2010. "Cooper's Hawk (*Accipiter cooperii*)." In *Raptors of New Mexico*, edited by J.-L. E. Cartron, 177–93. Albuquerque: University of New Mexico Press.

37. See chapter 12.

38. Courser, W. D. 1979. "Continued Breeding Range Expansion of the Burrowing Owl in Florida." *American Birds* 33:143–44.

39. Thomsen, L. 1971. "Behavior and Ecology of Burrowing Owls on Oakland Municipal Airport." *Condor* 73:177–92.

40. Poulin, R. G., L. D. Todd, E. A. Haug, B. A. Millsap, and M. S. Martell. 2011. "Burrowing Owl (*Speotyto cunicularia*)." In *The Birds of North America*, edited by P. G. Rodewald. Ithaca: Cornell Lab of Ornithology. Accessed March 9, 2017. https://birdsna.org/Species-Account/bna/species/burowl.

41. Millsap, B. A., and C. Bear. 2000. "Density and Reproduction of Burrowing Owls along an Urban Development Gradient." *Journal of Wildlife Management* 64:33–41.

42. Trulio, L. A. 1995. "Passive Relocation: A Method to Preserve Burrowing Owls on Disturbed Sites." *Journal of Field Ornithology* 66:99–106.

43. DeSante, D. F., E. D. Ruhlen, and R. Scalf. 2007. "The Distribution and Relative Abundance of Burrowing Owls in California during 1991–1993: Evidence for a Declining Population and Thoughts on Its Conservation." In *Bird Populations Monograph No. 1*, edited by J. H. Barclay, K. W. Hunting, J. L. Lincer, J. Linthicum, and T. A. Roberts, 1–41. Proceedings of the California Burrowing Owl Symposium, November 2003. Point Reyes Station, CA: The Institute for Bird Populations and Albion Environmental, Inc.

44. Wilkerson, R. L., and R. B. Siegel. 2010. "Assessing Changes in the Distribution and Abundance of Burrowing Owls in California, 1993–2007." *Bird Populations* 10:1–36.

45. Marzluff, J. 2016. "A Decadal Review of Urban Ornithology and a Prospectus for the Future." *Ibis* 159:1–13.

CHAPTER 19

Perspectives and Future Directions

Stephen DeStefano and Clint W. Boal

URBANIZATION IS BOTH A LANDSCAPE process and a demographic process. Most people are familiar with the changes on the landscape that have occurred on both local and global scales, and virtually all of us have experience with the profound changes in the neighborhoods, towns, and cities where we have lived. Such large-scale urbanization has occurred in all developed and most developing nations. In addition to these widespread landscape changes, demographic changes to the world's human population have reached the point where now more than half of the 7.3 billion people on the planet are urban dwellers.[1] This trend toward urbanization is projected to continue, with increasingly fewer people living and working in rural environments. Along with these changes to our demographic profile come different experiences, attitudes, and opinions about how we view our place in nature.[2]

Within the realm of ecological science, the subdiscipline of urban ecology has grown remarkably in the past few decades. Much has been written about the urban environment, including definitions of *urban*, *suburban*, and *exurban*; descriptions of urban wildlife habitat and habitat features for a variety of wildlife species; discussions of urban-rural gradients and heat sinks; and many other terms and concepts. One need only search the Internet for such terms to find a large volume of published and unpublished information on these and related

273

topics. In fact, a recent casual search for *urban ecology* in a popular scholarly search engine yielded more than two million results. It is clear that urbanization is of great interest to ecologists and conservationists and is considered a major driver of change on earth.

Likewise, ecological processes within urban environments continue to be explored and discussed online and in published articles. For example, so-called recombinant biological communities—that is, combinations of indigenous and many nonindigenous organisms, which heretofore had never lived or inter-acted together—make up the assemblages of plant and animal species that now live closest to humans.[3,4] Members of these relatively newly formed biological communities interact with people—and their products and associations—as a daily part of existence. Human subsidies, such as food, water, and shelter, and hazards—whether they be in the form of our pets, our automobiles, our chemi-cals, or our behaviors—make up the new reality for urban and suburban wildlife populations.

Such is the world of urban birds of prey, many species of which are now firmly entrenched in the urban and suburban world. This volume—on birds of prey in an urbanizing world—is both a synthesis of, and a testament to, the success that many species of raptors are having in human-dominated environ-ments. Research results and natural history accounts on raptors in urbanizing environments have taken their place in both the scientific and popular literature, both of which are now burgeoning with such articles. Many of these ideas are covered in the background, species accounts, and conservation and management chapters of this volume. Drawing on this information, we will attempt to provide a broader perspective on birds of prey in an urbanizing world and speculate on what the future might hold for their ecology and conservation on a rapidly changing planet.

The Urbanization of Wildlife

The degree to which an individual species is capable of existing in urban or suburban environments is constantly shifting for many species. Wildlife species that just a few decades ago were thought to require large patches of undeveloped, wild, and natural landscapes, devoid of human development with a minimum of human contact, are now part of urban-suburban settings. In the northeastern United States, wild turkeys (*Meleagris gallopavo*), American fisher (*Martes pen-nanti*), moose (*Alces americanus*), and beavers (*Castor canadensis*) now occupy landscapes fragmented by housing, businesses, roads, and other development and infrastructure, while javelinas (*Pecari tajacu*) and greater roadrunners (*Geococcyx californianus*) have wandered well into the interior of major cities in the Southwest.[5] Mountain lions (*Puma concolor*) prowl developed canyons in the West;[6] wolves (*Canis lupus*) venture into the suburbs of Anchorage, Alaska, to hunt moose; and coyotes (*Canis latrans*) can be found in virtually every major city in the United States.[7] So it is for many birds of prey, as discussed in this volume; species such as the Cooper's hawk (*Accipiter cooperii*), peregrine falcon (*Falco peregrinus*), and bald eagle (*Haliaeetus leucocephalus*), all once thought to require wild and remote places, not only occupy but flourish in cities today.

There are several reasons for this "evolution" in habitat and landscape use, which now includes urbanized areas for many raptor species. As R. William Mannan and Robert J. Steidl discussed in chapter 4, development may occur in already occupied habitat, where humans have essentially come to the wildlife. In other instances, some species move into urban and suburban areas to use resources, particularly food but also water and shelter, and to escape from predation or competition (color plate 14).[8] Whatever the reasons, we are seeing the exploitation of human development by some species, while others do not appear able to cope with these kinds of conditions and become more rare, or locally extirpated, as human development spreads across the landscape. This is a conundrum of modern conservation: urbanization can provide new opportunities for many different species, and yet at the same time, it negatively affects a host of other species, leading to poor demographic performance, local rarity, and regional extirpation.

This phenomenon of wildlife adapting to urban settings, however, also comes with some challenges for wildlife management. The adage that "wildlife management is really people management" has never been more appropriate than today, and this may be especially true for conserving and managing raptor populations in urban environments. As Brian E. Washburn detailed in chapter 15, human-wildlife interactions in urban settings are increasingly complex, the landscape of stakeholders is increasingly diversified, and stakeholder desire to take part in decisions is very high. In particular, issues related to human health, safety, and tolerances are critical considerations. However, many of these issues can be resolved by simply educating and working with the public; in chapter 17, John M. Davis provided several illustrative examples of how urban wildlife biologists have worked with people not only to remedy situations involving conflict but also to satisfy people in cases of interest and concern about raptors. This latter point is a notable and well-founded concern among the public. Wherever they occur, raptors have lives full of risk. Much of this is due to a predatory lifestyle and includes potential collisions while chasing prey, injuries from battle while subduing prey, becoming diseased or exposed to contaminants from prey, and even becoming prey themselves. However, as James F. Dwyer and his coauthors explained in chapter 14, urban settings can increase and exacerbate many of these risks compared to natural landscapes, especially issues such as collisions and electrocution. The public interest and empathetic concern for the well-being of urban raptors, which have led to monumental efforts to help them toward recovery and release, is highlighted by Lori R. Arent and her coauthors in chapter 16.

The Urban Raptor

Any raptor species may be occasionally seen moving over or through an urban landscape, especially during migration periods. However, in terms of actual "use" of the urban environments, there are variable levels of occupancy. At the most basic level are species that do not actually dwell in urbanized areas but may occasionally use areas within or adjacent to cities and towns. Some examples are rough-legged hawks (*Buteo lagopus*) and short-eared owls (*Asio flammeus*), both open-country species that would not typically enter a structurally diverse urban area but may inhabit adjacent open marshland edges, landfills, or grasslands created in association with airports (figure 19.1). More relevant to our concerns,

Figure 19.1. Many open-country species, such as this short-eared owl, rarely venture into heavily urbanized areas but may readily frequent adjacent open areas, including marshlands, airports, and fields like this one near Boston, Massachusetts. Photograph courtesy of Brian Rusnica.

however, are those species that regularly occupy urban settings, whether during the breeding season such as Mississippi kites (*Ictinia mississippiensis*) in North America (figure 19.2), primarily during winter such as sharp-shinned hawks (*Accipiter striatus*), or all year round such as red-shouldered hawks (*Buteo lineatus*) in North America, powerful owls (*Ninox strenua*) in Australia, and black kites (*Milvus migrans*) in Asia.

Before further considering the future of urban raptors, both in terms of conservation and research, it is prudent to briefly review and consider what we do know. There are numerous potential benefits and risks of living in urban landscapes for raptors,[8,9] many of which have been examined in depth within this volume. However, we can identify some generalities among those species that habituate and even flourish in urban settings. First, we should consider different contexts of urban use and how they relate to individual raptor species, their habitat choices, and their behavioral flexibilities.

Urban landscapes certainly differ from the adjacent rural landscapes, but as R. William Mannan and Robert J. Steidl pointed out in chapter 4, these differences can be dramatic or quite subtle depending on the region, and there are

Figure 19.2. Mississippi kites have adapted to human activities and expanded from their historic breeding range in the natural woodlands of the southeastern United States to now appear across the southern Great Plains, primarily in urban areas. Photograph courtesy of Clint W. Boal.

usually gradations from urban to rural. Urban areas often contain anthropogenic structures and tall trees as part of landscaping and urban forestry. Depending on location, this may be largely reminiscent of the previous natural landscape. However, urbanization may also create "islands" of urban forests in areas that once lacked trees. In chapter 2, Cheryl R. Dykstra pointed out that different species may be accepting of different zones along the urban-to-rural gradient of human-altered habitats, explaining that a primary factor influencing urban area use is the behavioral flexibility of a given species in habitat and nest-site selection, foraging habits, and other behaviors. Expanding on this, in chapter 3, Clint W. Boal demonstrated that some characteristics appear relatively consistent among those raptor species that regularly occupy urban settings. Primary among traits suggesting an urban presence is an inherent wide breadth of prey, which may indicate a flexibility that would be advantageous in an urban environment with potentially novel prey. Second, evidence suggests many urban raptors either predominantly prey on birds or adjust their food habits to take advantage of high numbers of avian prey in urban settings. Along with this diet breadth is a pattern in which urban raptors tend to shift their diet toward avian prey (figure 19.3). Finally, there is a tendency for urban raptors to be those for which the urban setting either merges with the adjacent "natural" habitat or in some way replicates "natural" habitat in areas where it does not normally occur.[10] For example, Robert N. Rosenfield and his coauthors noted in chapter 7 that Cooper's hawks in Stevens Point, Wisconsin, may readily use urban areas built within, and retaining parts of, existing woodlands, whereas Cooper's hawks in Tucson, Arizona, use created habitat associated with urbanization.

It is intuitive that urban-dwelling raptors require some level of adaptability or plasticity to acclimate to human development.[11,12] However, we cannot with certainty say that species x, because it occupies urban areas, is more adaptable than species y, which does not. For example, red-shouldered hawks are a woodland species and common in urban areas, but aplomado falcons (*Falco femoralis*) are an open grassland species not known to occupy urban areas. This does not mean that aplomado falcons are less adaptable than red-shouldered hawks. Rather than the aplomado falcon being "behaviorally inflexible," which falconers experienced with training the species would argue is not the case, it is more likely a question of what the species perceives as suitable habitat; a wooded urban landscape is nothing like the grasslands to which aplomado falcons are adapted. Alternatively, consider the ecologically similar peregrine falcon and prairie falcon (*Falco mexicanus*). Peregrines are quite common as urban residents, whereas prairie falcons are rarely found in urban areas. In this comparison, it could be

Figure 19.3. *A*, Cooper's hawks are primarily avian predators and find abundant resources in urban settings, Photograph courtesy of Doris Evans; *B*, Barred owls prey primarily on small mammals in exurban areas, but birds account for more than half their diet in urban settings, Photograph courtesy of Helen Reidel; *C*, Even in urban areas, Mississippi kites are primarily insectivorous but will take advantage of the abundance of avian prey when drought reduces invertebrate availability, Photograph courtesy of Clint W. Boal.

hypothesized that the peregrine falcon is more behaviorally adaptable than the prairie falcon, but differences in their use of urban areas could still be attributable to some differences in the natural habitat where they normally occur. Understanding what drives differences in adaptability remains elusive. This is one of several intriguing questions for future research efforts focused on the ecology and conservation of urban raptors.

Directions for Research

Foundational to our understanding of one component of urban ecology—urban raptors—is an evaluation of what we know and what we do not know about raptor biology, especially within the context of urban environments. Urban environments can provide a rich research setting for a closer examination of species' natural history, ecological relationships, and human-raptor interactions. For conservation, urban areas can be used or developed as wildlife habitat or could serve as buffers to environmental change (including climate change) for some species. Public involvement in a wide variety of interests is important for garnering support and avoiding or mitigating conflicts. Citizen scientists can be deployed to enhance information gathering and to engage the public. Finally, public safety considerations, such as careful management of large trees and snags that are important to many raptors species or their prey, may be particularly important in urban settings.

First, the majority of research on urban raptors has focused on nesting habitat and productivity. This is understandable, as the reliable presence of the species at a nest site provides some of the most readily available ecological information to collect on raptors (color plate 15). Less is known about survival, dispersal, wintering ecology, home range, or community ecology, though these topics are increasingly being studied for a few species. The very presence of raptors is one possible indicator of the potential success of some species in urban areas. However, presence alone does not tell the whole story; demographic performance—how well raptors survive and reproduce and the resulting trend in abundance or density over time—is the real key to measuring success or failure. Although a few studies have addressed some aspects of demographic performance in urban raptor populations, demographic data are challenging to collect for any wildlife species, requiring an abundance of time, effort, and resources. Hence information on demographic performance is relatively rare in the published literature, but it is vital for understanding the status and trends of urban raptor populations. Further, there are few studies examining differences in demographic performance between raptor species in urban and exurban areas.

Second, the majority of information on urban raptors is for diurnal species. Locating and monitoring nocturnal species (primarily owls) in urban environments introduces a suite of challenges, ranging from safety concerns related to working at night in some areas to delays created by concerned citizens who contact law enforcement about "suspicious activity," especially in residential areas (figure 19.4). Although nests may be easy enough to visit during daylight, owls are generally roosting and inactive during diurnal periods.

Third, there are very limited data to compare diet in urban and exurban areas for many species. Existing data for comparison are often from different regions or collected in different years; data that are spatially and/or temporally disjunct can result in misleading interpretations. Interestingly, there are probably better food habits data for nocturnal species than for most diurnal species because owls use consistent roosts and regurgitate easily identifiable skeletal

Figure 19.4. There is a substantial lack of information about owls in urban landscapes due to their nocturnal behaviors. Although productivity and food habits data may be obtained once nests are found, information on other aspects of life history are challenging to acquire. For example, the occurrence of this Mexican spotted owl in an urban backyard in Albuquerque, New Mexico, is quite unexpected. Photograph courtesy of Katelyn Bird.

material from their prey, but there is still a lack of comparable behavior data between urban and exurban locations. For example, increased abundance of prey in urban areas may allow some raptors to spend more time close to the nest protecting the young, thereby increasing reproductive success compared to exurban birds. Alternatively, disturbances near urban nests may reduce nest attentiveness or disrupt feeding and ultimately result in lower reproductive success compared to exurban birds. Comparative research on food habits and behaviors at both urban and exurban areas is important for understanding urban ecology and incorporating urban landscapes into conservation and management planning.

Fourth, few in-depth ecological studies have been conducted on urban raptors, and those that have are generally for single locations. There are notable exceptions, such as the species accounts in this volume that describe studies of peregrine falcons, Cooper's hawks, and red-shouldered hawks, which were conducted by multiple researchers in cities across a wide range of latitudes and longitudes. For widely distributed species, studies across multiple and well-spaced geographical areas would elucidate the range of conditions, and thus the potential conservation opportunities, that urban areas provide. Additionally, in concert with understanding which species can be successful in urban areas, it is just as important—indeed, perhaps more so—to understand which ones will not. Conservation efforts incorporating urban settings may work well for the former (e.g., peregrine falcons), but not the latter (e.g., prairie falcons) species.

The research opportunities urban settings provide to expand our understanding of basic biology and applied conservation of raptors are manifest. The general approachability of these often-secretive species and their presence in accessible locations provide ample opportunities to address questions on aspects of behavior including parental care, foraging ecology, and intra- and interspecies interactions. For example, how do larger predators, such as great horned owls (*Bubo virginianus*), influence where smaller raptors, such as Cooper's hawks or barred owls, choose to nest within the urban landscape (color plate 16)? For conservation purposes, how do different spatial characteristics and compositions of urban landscapes—such as stands of trees of different ages; proportion and extent of open spaces; and structures of different green spaces like golf courses, cemeteries, and school fields—influence raptor species' presence or absence? Additionally, how may these spaces function as high quality habitat for some species, marginal habitat for some, and population sinks for others? Questions related to basic biology and applied conservation are seemingly endless, but the urban environment can provide the means and circumstances to begin to answer some of these questions and to better comprehend the raptors' world.

Opportunities and an Eye to the Future

Urban raptors, as with all urban wildlife, are subjected to the dynamic and rapidly changing characteristics of urban ecosystems. Urban areas continue to grow across the landscape in many ways, including *sprawl* (rapid growth across the landscape), *conurbation* (the merging of urban areas through population growth and physical expansion), and *infilling* (buildup on undeveloped parcels within an urban area).[5,13] In addition, the makeup of urban environments presents heightened challenges, such as increased traffic volume and nonpermeable surfaces that alter temperature regimes and hydrology, as well as pollution, noise, light, and other characteristics that can affect biotic communities. How various raptor species adapt to these changing conditions remains to be seen. Furthermore, not all urban areas are the same, and differences among ecoregions—among other variables—can affect habitat availability and relationships between urban and adjacent nonurban areas. A metropolitan area in the Northeast differs significantly from a city in the southwestern United States, and these different characteristics need to be considered.

Climate change is predominant among potential current and future drivers of environmental change.[14] Urban centers are among the areas most influenced by climate change, in part because of their role as heat islands in which buildings and roads absorb heat during the day and release it slowly at night. Alternatively, urban centers might become oases for some species because of the increased availability of resources such as water and cover. Climate change also affects the distribution and prevalence of diseases and parasites in many parts of the world; species of wildlife that were once very abundant have declined precipitously because of changes in diseases and parasites brought on by climate change and other factors.[15,16] These dynamics alter the circumstances faced by biotic communities within urban environments. As the world's climate changes, there will be winners and losers among the members of an ecological community. For raptors, it is not clear which will thrive and which will not, as climate and local conditions change over time; climate change is obviously an important area of research and concern for raptor biologists.

The human element is always a critical factor in conservation, and most biologists recognize the importance of human dimensions in raptor management and conservation. Stakeholder involvement is essential to the success of any management effort and in many cases the public now expects and demands to be kept informed, to offer opinions, and to be allowed to participate in any activities proposed by conservation agencies. People's attitudes toward certain wildlife species can also change. Animals that were once cherished can become pests, a change in perception that is often associated with a change in abundance.[17] Wild turkeys are a good example of this: they have become perceived as pests in some urban areas where they have become very abundant.[18] It may not be too farfetched to think that an abundant urban raptor may someday reach pest status among some segments of the human urban population. Elements of human health and safety are also of concern. For example, urban forest management places a high priority on management of hazardous trees.[19] Large trees with cavities and dead limbs and standing dead trees are important for many species of raptors and their prey. However, these can also present significant safety threats to people and property and need to be managed by professionals who understand the dynamics of tree structure and issues related to human safety.

Ultimately, environmental issues, including the conservation of biodiversity, are probably at the highest levels of public awareness in human history. Increasingly, people have an appreciation of the wide range of wildlife, from invertebrates to megamammals. Yet some species—among them apex predators, including many raptors—tend to capture the attention and imagination of many people more than other species. Their role as high trophic level predators in ecological communities, their attention-grabbing behaviors, and their power and grace in movement make them symbolic among wildlife in urban areas and provide an important connection between people and nature.

Acknowledgments

We thank our many colleagues and the discussions that have influenced our thoughts and perspectives on the intersection of wildlife and urban settings. We thank David Andersen and Todd Fuller for their reviews of this manuscript and their constructive comments. We are especially appreciative for the photographs donated by Katelyn Bird, Doris Evans, Kim Domina, Helen Riedel, and Brian Rusnica.

Literature Cited

1. United Nations. 2015. "World Population Prospects: The 2015 Revision, Key Findings and Advance Tables." Working Paper No. ESA/P/WP.241.

2. Freyfogle, E. T. 2017. *Our Oldest Task: Making Sense of Our Place in Nature.* Chicago: University of Chicago Press.

3. Soule, M. E. 1990. "The Onslaught of Alien Species, and Other Challenges in the Coming Decades." *Conservation Biology* 4:233–40.

4. Rotherham, I. D. 2017. *Recombinant Ecology—A Hybrid Future?* New York: Springer International Publishing.

5. DeStefano, S., and D. M. DeGraaf. 2003. "Exploring the Ecology of Suburban Wildlife." *Frontiers in Ecology and the Environment* 1:95–101.

6. Moss, W. E., M. W. Alldredge, and J. N. Pauli. 2016. "Quantifying Risk and Resource Use for a Large Carnivore in an Expanding Urban—Wildland Interface." *Journal of Applied Ecology* 53:371–78.

7. Weckel, M., and A. Wincorn. 2016. "Urban Conservation: The Northeastern Coyote as a Flagship Species." *Landscape and Urban Planning* 150:10–15.

8. Mannan, R. W., and C. W. Boal. 2004. "Birds of Prey in Urban Landscapes." In *People and Predators*, edited by N. Fascione, A. Delach, and M. E. Smith, 105–17. Washington, DC: Island Press.

9. Hager, S. B. 2009. "Human-Related Threats to Urban Raptors." *Journal of Raptor Research* 43:210–26.

10. Andersen, D. E., and D. L. Plumpton. 2000. "Urban Landscapes and Raptors: A Review of Factors Affecting Population Ecology." In *Raptors at Risk: Proceedings of the V World Conference on Birds of Prey and Owls*, edited by R. D. Chancellor and B.-U. Meyburg, 434–45. Berlin, Germany: World Working Group on Birds of Prey and Owls; Surrey, BC: Hancock House Publishers.

11. Steidl, R. J., and B. F. Powell. 2006. "Assessing the Effects of Human Activities on Wildlife." *George Wright Forum* 23:50–58.

12. Carrete, M., and J. L. Tella. 2011. "Inter-Individual Variability in Fear of Humans and Relative Brain Size of the Species Are Related to Contemporary Urban Invasion in Birds." *PLoS ONE* 6: e18859.

13. Ramalho, C. E., and R. J. Hobbs. "Time for a Change: Dynamic Urban Ecology." *Trends in Ecology and Evolution* 27:179–88.

14. NASA. 2017. "Global Climate Change: Vital Signs of the Planet." https://climate.nasa.gov.

15. Hassell, J. M., M. Begon, M. J. Ward, and E. M. Fèvre. 2017. "Urbanization and Disease Emergence: Dynamics at the Wildlife-Livestock-Human Interface." *Trends in Ecology and Evolution* 32:55–67.

16. Young, H. S., I. M. Parker, G. S. Gilbert, A. S. Guerra, and C. L. Nunn. 2017. "Introduced Species, Disease Ecology, and Biodiversity-Disease Relationships." *Trends in Ecology and Evolution* 32:41–54.

17. DeStefano, S., and R. D. Deblinger. 2005. "Wildlife as Valuable Natural Resources vs. Intolerable Pests: A Suburban Wildlife Management Model." *Urban Ecosystems* 8:179–90.

18. Miller, J. E., B. C. Tefft, R. E. Eriksen, and M. Gregonis. 2000. "Turkey Damage Survey: A Wildlife Success Story Becoming Another Wildlife Damage Problem." *Proceedings of the Wildlife Damage Management Conference* 9:24–32.

19. Kane, B., P. S. Warren, and S. B. Lerman. 2015. "A Broad Scale Analysis of Tree Risk, Mitigation and Potential Habitat for Cavity-Nesting Birds." *Urban Forestry and Urban Greening* 14:1137–46.

Contributors

Clifford M. Anderson is the founder of the Falcon Research Group based in Bow, Washington, which focuses on the conservation of birds of prey through education. His research interests include migration, breeding, wintering, and raptor DNA. His species of interest include the bald eagle and peregrine falcon.

Lori R. Arent is the clinic manager at the Raptor Center at the University of Minnesota. She is the author of *Raptors in Captivity: Guidelines for Care and Management*, which was adopted by the US Fish and Wildlife Service as their standard on captive raptor management.

John (Jack) H. Barclay retired in 2013 from Albion Environmental, Inc., an environmental consulting company that he cofounded in California in 1997. He has engaged in burrowing owl conservation policy and management practices and conducted burrowing owl research and management at a major international airport for more than 20 years. Before moving to California, he worked for 11 years for the Peregrine Fund at the Cornell Lab of Ornithology on the peregrine falcon reintroduction program in the eastern United States.

Douglas A. Bell is the wildlife program manager for the East Bay Regional Park District, where one of his duties involves assessing the effects of wind energy projects and other anthropogenic development on raptors, with an emphasis on golden eagles and prairie falcons. As a research associate with the California Academy of Sciences, he has been engaged in population genetic studies of falcons.

Richard O. (Rob) Bierregaard and his graduate students from UNC–Charlotte studied the thriving population of barred owls in and around Charlotte, NC, for 10 years. Rob is currently a research associate in the Ornithology Department at the Academy of Natural Sciences of Drexel University.

Keith L. Bildstein is the Sarkis Acopian director of conservation science at Hawk Mountain Sanctuary. He is the author and editor of numerous books about raptors, including *Raptors: The Curious Nature of Diurnal Birds of Prey*. His current projects include studies of striated caracaras on the Falkland Islands, long-term studies of new world vultures, and work on Africa's hooded vultures.

David M. Bird is emeritus professor of wildlife biology at McGill University. Over four decades of studying raptors, many of them in urban settings, he has written almost 200 peer-reviewed papers, supervised 50 graduate students, and written/edited a dozen books. He is currently one of the world leaders in applications of drone technology for birds and is the founding editor of the *Journal of Unmanned Vehicle Systems*.

Peter H. Bloom is the president and zoologist of Bloom Research Inc., and in that role, he focuses on long-term field studies that enhance the conservation of birds of prey and their habitats.

Clint W. Boal is a research and wildlife biologist with the US Geological Survey's Texas Cooperative Research Unit and holds a joint appointment as a professor of wildlife ecology at Texas Tech University. He has conducted research with birds of prey for over 25 years.

Gail Buhl is the education program manager at the Raptor Center at the University of Minnesota. She is responsible for the raptor education program and training and maintenance of the birds. She has more than 25 years of experience in raptor education and wildlife rehabilitation.

Courtney J. Conway is the leader of the US Geological Survey's Idaho Cooperative Fish and Wildlife Research Unit and is a professor in the Department of Fish and Wildlife Sciences at the University of Idaho. His research interests include behavioral ecology, life history evolution, migration, and effects of management actions on bird and mammal populations.

Raylene Cooke is an associate professor of wildlife and conservation biology in the School of Life and Environmental Sciences at Deakin University, Australia. Her research focuses on raptors, urban ecology, and the impact of urban gradients on biodiversity. She has been researching the impact of urbanization on powerful owls for more than 20 years.

John M. Davis is a former urban wildlife biologist with the Texas Parks and Wildlife Department who is now the agency's wildlife diversity program director. His program oversees the agency's urban wildlife program, nongame and rare species management and permitting, environmental review, citizen science, natural heritage data management, and the Texas Master Naturalist Program. His professional interests include connecting people to wildlife and promoting ecologically responsible development.

Edward Deal is a retired physical therapist with a long love of raptors. He has helped study and band urban-nesting peregrine falcons for 25 years and also coordinates a long-term study on urban-nesting Cooper's hawks in Seattle.

Stephen DeStefano is leader of the US Geological Survey's Massachusetts Cooperative Fish and Wildlife Research Unit and a research professor at the University of Massachusetts–Amherst. As a wildlife biologist, he has worked on a wide array of species, including waterfowl, raptors, carnivores, and ungulates. His research focuses on population ecology, wildlife-habitat relationships, and human-wildlife interactions.

James F. Dwyer is a research scientist at EDM International, where he develops and publishes original research to facilitate environmentally responsible operation of electric utilities. His main focus is on minimizing avian electrocutions and collisions on overhead power structures.

Cheryl R. Dykstra is an independent researcher and consultant. She serves as editor-in-chief of *The Journal of Raptor Research* and has spent over two decades leading raptor research projects, including an ongoing 20-year study of urban red-shouldered hawks.

Marcel A. Gahbauer is a senior wildlife ecologist at Stantec with more than 16 years of field and reporting experience involving species at risk, population monitoring, and environmental assessments. He is also executive director at the McGill Bird Observatory, a program that involves more than 5,000 birds and 100 people annually.

Sofi Hindmarch is a project biologist at the Fraser Valley Conservancy who has studied species at risk in the Fraser Valley for 10 years, with a focus on the barn owl. Since 2010, she has focused on conservation issues affecting barn owls, such as the loss of habitat and nest sites, road mortality, and the risk of secondary poisoning from rat poisons.

Fiona Hogan is a lecturer in conservation biology and director of WildDNA at Federation University, Australia. As a molecular ecologist, her research focuses on designing noninvasive sampling methodologies for applied population genetic studies of wild animals.

Bronwyn Isaac is an associate lecturer in biological sciences at Monash University, Clayton, Australia. Her research is centered on the ecology and conservation of raptors within altered environments and aims to produce biologically resilient systems that can coexist alongside human populations.

Lloyd Kiff has been past president of the Cooper Ornithological Society, has been on the board of directors of the Raptor Research Foundation, and is a fellow of the American Ornithologists' Union. His research interests include raptor conservation, zoology, host parasitism, contaminant effects, DDE-induced eggshell thinning, history of American ornithology, and constructing bibliographic databases. His species of interest are the California condor and the peregrine falcon.

Gail E. Kratz is the rehabilitation director for the Rocky Mountain Raptor Program (RMRP), where she is responsible for the medical management, food procurement, and housing of all active case raptors and permanent educational ambassadors. She has been with the RMRP since 1989.

Jeffrey L. Lincer is a past president of the Raptor Research Foundation and the Wildlife Society's Southern California Chapter and is a founding member of the Global Owl Project. He is most well known for his classic research on the effects of DDT and PCBs on raptor eggshell thickness and his leadership role in the conservation and management of numerous raptor species, especially the burrowing owl.

R. William Mannan is a professor of wildlife, fisheries, and conservation biology at the University of Arizona. His research focuses on relationships between animals and their habitats in urban and forest environments and animal behavior related to habitat use. He also leads a long-term project examining the dynamics of a population of urban-nesting Cooper's hawks.

Michael D. McCrary is an associate of Bloom Research Inc. His main interest is movement ecology of raptors, including migration and natal and breeding dispersal.

F. Arthur McMorris is the peregrine falcon coordinator of the Pennsylvania Game Commission. He also participates in numerous other bird population studies and conservation projects.

Brian A. Millsap is the national raptor coordinator for the US Fish and Wildlife Service. He has conducted field and museum research on a number of raptor species for more than 35 years. His research interests include population ecology, raptor demography, taxonomy, and conservation. His species of interest are the Cooper's hawk, bald eagle, gray hawk, golden eagle, peregrine falcon, and burrowing owl.

Joel E. Pagel is a raptor ecologist for the US Fish and Wildlife Service's National Raptor Program, where he conducts research and applied management on peregrine falcons, golden eagles, and other raptors. He spent the first two-thirds of his 34-year government career with the US Forest Service as the agency's first peregrine falcon specialist.

Patrick T. Redig is a professor at the University of Minnesota and founder and honorary director of the world-renowned Raptor Center. He has been responsible for many advances in avian orthopedic surgery and raptor medicine.

Robert N. Rosenfield is a professor of biology at the University of Wisconsin–Stevens Point. He has conducted population and behavioral ecology studies of breeding Cooper's hawks in Wisconsin for 38 years. He has also conducted research on raptor migration and nesting peregrine falcons in Greenland.

Robert Sallinger is the conservation director for the Audubon Society of Portland and an adjunct professor of law at Lewis and Clark Law School. His current responsibilities for Portland Audubon include directing local, regional, and national conservation policy initiatives, wildlife research initiatives, the Backyard Habitat Certification Program, and the Wildlife Care Center.

Greg Septon has directed and managed an urban peregrine falcon recovery effort in Wisconsin for nearly 30 years and has banded nearly 900 wild-produced peregrines. In 2014, he received the Noel J. Cutright Conservation Award from the Wisconsin Society for Ornithology for his work with endangered and threatened species.

Ben R. Skipper is an assistant professor of biology at Angelo State University. His research is focused on the distribution and life history of neotropical birds with emphasis on birds of prey. He serves as the curator of ornithology at the Angelo State University Natural History Collections.

Robert J. Steidl is a professor of wildlife and fisheries science at the University of Arizona. His research focuses on the effects of human activities on populations of animals and finding ways to mitigate those effects.

Jean-François Therrien is senior research biologist at Hawk Mountain Sanctuary. His current projects include studies of arctic raptors, including snowy owls, rough-legged hawks, and peregrine falcons, as well as several temperate species such as kestrels and vultures.

Brian E. Washburn is a research biologist with the USDA's APHIS Wildlife Services at the National Wildlife Research Center. His research involves basic and applied wildlife ecology studies that provide a better understanding of wildlife movement patterns, foraging ecology, habitat management, land use practices, and ecology of wildlife within urban ecosystems.

Marian Weaving earned her PhD in environmental science from Deakin University. Her research interests are focused on the ecology and behavior of endemic nocturnal species like the tawny frogmouth and its response to urbanization in Melbourne, Australia.

John G. White is an associate professor in wildlife and conservation biology in the School of Life and Environmental Sciences at Deakin University, Australia. His research interests focus on how species and communities respond to disturbance processes such as urbanization and fire. His current research includes understanding the ecology of urban powerful owls, and he also has a long-term program investigating the roles of fire and climate on small mammal communities.

Michelle Willette is an assistant professor and staff veterinarian of the Raptor Center, University of Minnesota, where she oversees the medical and surgical care of more than 800 raptors a year. Her main research interest is the development of the Clinical Wildlife Health Initiative, an interdisciplinary collaborative formed to promote the infrastructure and analytical tools required to utilize data from wildlife in rehabilitation settings.

Index

Page numbers followed by *f* and *t* refer to figures and tables, respectively.

abattoirs, 10, 11
abundance
  habitat, 20, 37, 54, 79–80, 128, 259–60
  food (*see* food: availability of)
  raptors, 9, 25, 37–38, 55–56, 79, 94, 114,
    174, 280, 283–84
accipiters, 95, 98, 126, 249
  nests, 25
  See also individual species
adaptability, 18–19, 113, 183–84, 274, 275, 278
aggression, 28, 97, 217–18, 222–23, 238, 250–
  51, 268. *See also under individual species*
airports, 174, 201, 218, 222, 276
air strikes. *See* collisions: aircraft
Andrus, Cecil, 4
antiperching devices, 223
Argentina, 11, 26, 28, 126, 173
Arizona, 55, 57, 80–82, 85, 101, 126, 128, 130–
  32, 134, 169, 207. *See also* Tucson, Arizona
ARs. *See* rodenticides: anticoagulant
Association of Zoos and Aquariums, 239
Audubon, John, 11
Australia, 152, 154, 156, 161, 277. *See also*
  Melbourne, Australia
AZA. *See* Association of Zoos and Aquariums

badger, American (*Taxidea taxus*), 169
Bald and Golden Eagle Protection Act, 221, 240
banding, 83, 84, 112*f*, 119–20, 145, 160, 182,
  183*f*, 187, 189*f*, 190*f*, 237, 262

bats, 9, 82
BBL. *See* Bird Banding Laboratory
BBS. *See* Breeding Bird Survey
beaver (*Castor canadensis*), 274
beech (*Fagus grandifolia*), 113
bioaccumulation, 203
biological control, 216
Bird Banding Laboratory, 84, 237
bird feeders, 13, 18, 19, 26, 27, 130, 201
  as conflict source, 207, 238, 249, 259
  as disease vector, 264
  as research subject, 263, 268
bird strikes. *See* collisions: aircraft
bird-watching, 215–16
Blackland Prairie Raptor Center, 254
blackwood (*Acacia melanoxylon*), 155
Breeding Bird Survey, 94
British Columbia, 26, 44, 56, 94, 99, 138–40.
  See also Victoria, British Columbia
buffer zones, 174, 266, 268
buteos, 126
  nests, 25
  See also individual species

cactus, 130, 131. *See also* saguaro
California, 21*t*, 24, 55, 57–58, 66, 99, 115*f*, 168,
  169, 171, 174, 186, 220, 265–67
  central, 25, 101, 114, 118
  southern, 20, 22, 25, 101, 114–15, 118–21
California Environmental Quality Act, 267

call playback, 156
caracara, crested (*Caracara cheriway*), 41*t*, 200
Carolina Raptor Center, 146
casuarina, tall (*Casuarina littoralis*), 155
CEQA. *See* California Environmental Quality
    Act
Charlotte, North Carolina, 21*t*, 25, 138–48,
    141*f*, 143*t*
cherry, native (*Exocarpos cupressiformis*), 155
chimneys, 140, 233*f*
chipmunk, eastern (*Tamias striatus*), 93
cicadas (Cicadidae spp.), 81
Cincinnati, Ohio 21*t*, 25, 112*f*, 113–14, 116,
    117, 118, 139, 140, 147
Circuitscape, 69
citizen science, 38–39, 235, 280
climate change, 103, 280, 283
Clinical Wildlife Health Initiative, 236
collisions, 59, 84, 98, 118, 131, 232
    aircraft, 12, 201, 218–19, 218*f*, 221, 223–24
    mitigation of, 201
    power line, 52, 119, 201
    vehicle, 19, 52, 119, 146, 148, 173, 200–201,
        233*f*, 254
    window, 52, 201, 202*f*, 207, 268
    wire, 201
Columbidae, 59. *See also individual species*
commensalism, 10–11, 65. *See also* symbiosis
computerization, 67
condor, Andean (*Vultur gryphus*), 12
conifers, 95. *See also* pine; *individual species*
conjunctivitis, 264
connectivity, 69
conservation, 240–42, 267–69, 274, 280, 282–84
    funding, 246–47
    of habitat, 67, 269, 280
    networking, 254, 256
    outreach, 240–42, 243
    publicity, 182
    resources, 246–47, 256
    and technology, 156–57
    and wildlife communities, 37
    See also under individual species
conspecifics, 21, 44, 46
contaminants, 94, 203–4, 216, 232, 276. *See
    also* pesticides; poison; rodenticides
conurbation, 283
Cornell Lab of Ornithology, 38
cottonwood (*Populus* spp.), 80–81
cottonwood, Fremont (*Populus fremontii*), 114
coyote (*Canis latrans*), 131, 275
creosote bush (*Larrea tridentata*), 56

cross-fostering, 58
crow, American (*Corvus brachyrhynchos*), 140
crow, pied (*Corvus albus*), 10
CWHI. *See* Clinical Wildlife Health Initiative
cypress, bald (*Taxodium distichum*), 139

Darwin, Charles, 11
DDE, 94, 180–81, 203
DDT, 58, 94, 180, 203
degu (*Octodon degus*), 130
demographics, 19, 51–55, 69, 98, 280
    and urbanization, 52*f*
depredation, 37
deserts, 56, 58
dieldrin/aldrin, 94, 181
diet, 26–28, 280. *See also* food; prey; *see also
    under individual species*
disease, 131, 204–6, 232, 264, 276
    zoonotic, 205, 236
    See also individual diseases
disking, 265
dispersal, 68, 69, 120, 132, 134, 146, 161. *See
    also under individual species*
distribution, 38–42, 39*t*, 41*t*
    and food, 42, 44*t*
disturbance, 80, 95, 135, 155, 188, 267, 282
    tolerance, 65, 116 (*see also under individual
        species*)
diversity, 18–19, 37
    genetic, 78
    in habitat, 23, 56–57, 120, 139
    of prey, 260
DNA, 67, 160
DNA profiling. *See* genetic profiling
Doppler, 237
dove, mourning (*Zenaida macroura*), 93
dove, Pacific (*Zenaida meloda*), 130
dove, rock. *See* pigeon, rock
dragonflies (Odonata spp.), 82
drought, 121
drowning, 131
dumps, 10, 27. *See also* garbage; landfills

eagle, bald (*Haliaeetus leucocephalus*), 40, 41*t*,
    206–7, 254, 275
eagle, golden (*Aquila chrysaetos*), 41*t*, 206*f*
eagle, Spanish imperial (*Aquila adalberti*), 29
eagles
    exhibition, 240
    rehabilitation, 230
    See also individual species
eBird, 38–39, 40

eco-geographical variables, 156–57
ecological equivalents, 146–47
ecological niches, 37–38
ecological sinks, 146, 148
ecological traps, 27, 59, 101–2, 158, 161, 169
ecosystem services, 216
ecotourism, 216
education, 89, 121, 182, 188, 190–91, 221, 243,
    255, 262, 268
  and conflict resolution, 248
  effectiveness, 241
  environmental, 216, 238–39
  federal, 240
  formal programs, 239
  legal requirements, 239
  live presentations, 239–41, 242
  online, 242
  professional, 237–38, 255
  See also outreach
effigies, 89, 90, 255
EGV. *See* eco-geographical variables
electrocution, 19, 59, 118, 119, 131–32, 132*f*,
    133*f*, 136, 201–3
Emlen, Steve, 36
Endangered Species Act, 230
entrapment, 131, 207, 232, 252
environment
  abiotic, 53
  biotic, 53
  conversion, 19, 56, 168, 265, 266, 283
  stability, 54
  See also habitat; landscapes
EPC. *See* extra-pair copulation
EPF. *See* extra-pair fertilization
estimate of urban use, 42–43
Eucalyptus spp., 56, 114, 128, 134, 155, 206
euthanasia, 234, 235*t*
EUU. *See* estimate of urban use
evolution, 13, 59, 95
Excellence in Environmental Education:
    Guidelines for Learning, 239
extra-pair copulation, 28–29, 86
extra-pair fertilization, 68, 97, 160
extrapaternity, 97

failure to thrive, 232
falcon, aplomado (*Falco femoralis*), 278
falcon, bat (*Falco rufigularis*), 6
falcon, peregrine (*Falco peregrinus*), 3, 41*t*, 58,
    180, 183*f*, 275
  adaptability, 278–79
  aggression, 188, 189*f*, 258, 262

defense, 188, 262
diet, 6, 9, 12, 26, 66, 182, 260
and education, 187
habitat, 181
hacking, 5, 52, 182, 192*f*
human conflicts, 188, 191
imprinting, 5, 8
mortality, 188, 260–61
nests, 24, 53–54, 182, 183–84, 186, 186*f*,
    189*f*, 190*f*, 191*f*, 192*f*, 193*f*, 260–62
productivity, 58
reintroduction, 181–83
reproduction, 182
survival, 58
falcon, prairie (*Falco mexicanus*), 41*t*, 42,
    278–79
falconry, xii, 5, 135, 189, 240
falcons
  as animal control, 3
  nests, 24
  See also individual species
feathers, and DNA, 67, 160
Federal Aid in Wildlife Restoration Act, 247
feeding
  courtship, 97
  by humans, 10 (*see also* bird feeders)
fidelity
  mate, 160 (*see also* monogamy)
  nest, 89, 160, 174
  site, 82–83, 174
fir, Douglas (*Pseudotsuga menziesii*), 95
Fish and Wildlife Service, 255
fisher, American (*Martes pennanti*), 274
flame retardants, 203–4, 264
flexibility. *See* adaptability
flight initiation distance, 28
Florida, 55, 94, 98, 113, 121, 138–39, 169, 170,
    172, 174, 181, 265–66
flycatcher, scissor-tailed (*Tyrannus forficatus*),
    134
FNB. *See* food niche breadth
food
  availability of, 9, 18–19, 22, 26–27, 29, 45–
    46, 55–59, 66, 81–82, 98, 121, 130, 154,
    169–71, 222, 259–60, 263, 282
  and raptor communities, 38
  scavenged, 27
  See also diet; feeding; prey
food chain, 26, 203
food niche breadth, 43–45, 45*t*
forests, 57–58, 145, 207, 278, 284. *See also*
    habitat: forest

fostering, 58, 182
FRAGSTATS, 66
FWS. *See* Fish and Wildlife Service

garbage, 6, 9–11. *See also* dumps; landfills
Gehlbach, Frederick, 54
generalists, 12, 26, 27, 65, 110–11, 142, 200
genetic profiling, 160–61
genetic tags, 160–61
Geographical Information System, 66, 69, 157
Geospatial Modeling Environment, 66
GIS. *See* Geographical Information System
global positioning system, 67, 69, 158, 237
global system for mobile communication, 237
GME. *See* Geospatial Modeling Environment
goshawk, northern (*Accipiter gentilis*), 21t, 41t, 42, 47, 51, 200
    diet, 22, 26–27
    habitat, 20, 22
    nests, 25
    ranges, 21t, 22
GPS. *See* global positioning system
grasslands, 58
Great Plains, 36–37, 40, 79–83, 85, 94
ground squirrel, California (*Otospermophilus beecheyi*), 169
ground squirrel, round-tailed (*Xerospermophilus tereticaudus*), 169
GSM. *See* global system for mobile communication

habitat, 20–22, 21t, 51–60
    availability, 19, 21, 56, 60, 99, 140, 169, 283
    and conservation, 67, 269, 280
    desert, 53, 57, 127–28, 130–31, 134
    forest, 23, 47, 59, 66, 80–82, 94–95, 113–14, 138–39, 142, 145–49, 154–55, 157–58, 213
    fragmentation of, 19–20, 36, 56, 65, 68–69, 95, 154
    grassland, 53, 56, 113, 128, 170, 172, 222, 276, 278
    high-quality, 52, 54, 101–2
    management, 223
    and mortality, 98, 173
    recognition of, 43, 44t
    residential, 113–14, 140
    riparian, 80–82, 111, 113, 114, 157
    selection of, 4, 38, 80–81, 112–16
    suitability, 98, 157, 159
    swamp, 113, 139

    and water, 114, 127, 140, 181, 204–5
    wetland, 113
    woodland, 42, 46–47, 80, 94, 113–14, 128, 158, 213, 259, 278
    See also under individual species
habitat loss, 94, 135, 154, 155, 157, 232
habituation, 95, 171, 200
hackberry (*Celtis* spp.), 128
hacking, 5, 24, 52, 58, 182, 192f
harassment, 88. *See also* persecution
harrier, northern (*Circus hudsonius*), 41t, 42
hatch dates, 83
hawk, bay-winged. *See* hawk, Harris's
hawk, broad-winged (*Buteo platypterus*), 41t
hawk, common black (*Buteogallus anthracinus*), 41t, 42
hawk, Cooper's (*Accipiter cooperii*), 13, 21t, 40, 41t, 45t, 93, 96f, 134–35, 237, 249, 259, 263, 275, 279f
    aggression, 264
    defense, 28
    density, 57, 99, 100f
    diet, 27, 45t, 93
    distribution, 94, 100f
    habitat, 20, 102
    mortality, 59, 98–99
    nests, 23, 25, 57, 94–95, 101
    population, 94, 95
    ranges, 21t, 22, 57
    reproduction, 28–29, 68, 97, 99–101, 103
    survival, 57–58, 59, 98, 102
    tolerance, 66
hawk, ferruginous (*Buteo regalis*), 21t, 22, 41t, 42
hawk, gray (*Buteo plagiatus*), 41t
hawk, Harris's (*Parabuteo unicinctus*), 41t, 42, 126, 132f, 133, 207
    aggression, 131, 134–35
    conservation, 132, 135
    cooperative hunting, 130
    density, 134
    diet, 28, 129–30
    dispersal, 132, 134
    habitat, 128
    interactions, 134–35
    management, 135
    mortality, 59, 131
    nests, 128, 129f, 134
    predation, 131
    ranges, 129
    reproduction, 131
    sociality, 127

hawk, red-shouldered (*Buteo lineatus*), 21*t*,
  41*t*, 55, 110, 112*f*, 115*f*, 254, 277
  aggression, 117, 135
  and barred owls, 146–47
  conservation, 120–21
  defense, 28, 250
  diet, 28, 111, 116
  dispersal, 120
  distribution, 111*f*
  habitat, 20, 113
  mortality, 118–19
  nests, 23, 25, 111, 114, 116–17
  predation, 118–19
  ranges, 21*t*, 22, 111, 114, 115–16
  reproduction, 118
  survival, 119
hawk, red-shouldered, western (*Buteo lineatus
  elegans*), 114–15
hawk, red-tailed (*Buteo jamaicensis*), 3, 21*t*, 40,
  41*t*, 55, 135, 220, 259
  diet, 37
  nests, 25, 53–54
  ranges, 21*t*, 22
  as raptor predator, 101
hawk, rough-legged (*Buteo lagopus*), 41*t*, 42, 276
hawk, sharp-shinned (*Accipiter striatus*), 13,
  40, 41*t*, 42, 147, 249, 277
  nests, 24
hawk, Swainson's (*Buteo swainsoni*), 24, 25, 40,
  41*t*, 111
hawk, zone-tailed (*Buteo albonotatus*), 41*t*, 42
heat-island effect, 54, 283
HEOD. *See* dieldrin/aldrin
hickory (*Carya* spp.), 113
Houston, Texas, 248–49, 250, 255
HPAI. *See* influenza, avian
human activity, 24, 36, 102, 110, 135, 147
  intolerance, 20, 47, 87–88
  tolerance, 47, 51, 53, 60, 80, 128, 168, 171,
    267 (*see also under individual species*)
human-raptor interactions, 232, 243
  benefits, 215
  challenges, 217–21
  conflicts, 89, 188, 214–15, 239, 248–59
  economic impact, 221
  management, 221–25, 239, 248, 250–55
human-use areas, 20
human-wildlife interactions, 87, 275
hybridization, 148

imprinting, 8
inbreeding, 160, 187

Indiana, 20, 21*t*, 22, 98–99, 263
infilling, 283
influenza, avian (Orthomyxoviridae spp.),
  205
injury, 275
  anthropogenic, 230, 232, 254
insectivores, 8, 81
insects, 3, 8–9, 12, 28, 81, 86, 170–71
International Union for Conservation of
  Nature, 10–11
International Wildlife Rehabilitation Council,
  230
IUCN. *See* International Union for
  Conservation of Nature
IWRC. *See* International Wildlife
  Rehabilitation Council

javelinas (*Pecari tajacu*), 275

kangaroo rat, banner-tailed (*Dipodomys
  spectabilis*), 169
Kansas, 55, 80, 84–85, 88
kestrel, American (*Falco sparverius*), 40, 41*t*,
  134, 259
  diet, 26–27, 28
  nests, 24
kestrel, Eurasian (*Falco tinnunculus*), 44, 45*t*
kestrel, lesser (*Falco naumanni*), 3, 8
  diet, 12, 66
  nests, 24
  reproduction, 28–29
kestrels, 6. *See also individual species*
kingbird, western (*Tyrannus verticalis*), 134
kite, black (*Milvus migrans*), 6, 9, 10, 277
  diet, 27
kite, Mississippi (*Ictinia mississippiensis*), 28,
  41*t*, 42, 54–55, 86*f*, 88, 277, 279*f*
  abundance, 79
  aggression, 87–90, 88*f*, 135
  defense, 28
  density, 80, 82, 85
  diet, 37, 81–82, 85, 86
  habitat, 80–81
  management, 88
  mortality, 84
  nests, 23, 80–81, 82, 83, 277*f*
  predation, 86
  reproduction, 83–84
  success, 83
  survival, 84–85
  tolerance, 80
kite, red (*Milvus milvus*), 6, 7*f*

kite, swallow-tailed (*Elanoides forficatus*), 41*t*, 42
kite, white-tailed (*Elanus leucurus*), 28, 41*t*

landfills, 9, 203–4, 276. *See also* dumps
landscape genetics, 68–69
landscapes
    anthropogenic, 67–68, 116
    classification, 19
    connectivity, 68–69
    urban, 131
Last Chance Forever, 254
lethal removal, 225
light
    artificial, 9, 66, 68, 97, 169, 187
    moon, 97
    sun, 95
lion, mountain (*Puma concolor*), 275
live raptor presentations. *See* education: live presentations
livestock, 219
longevity, 84, 119, 145
Lubbock, Texas, 36–37, 83, 84, 85

maladaptive habitat choices, 157–58. *See also* ecological traps
mammals, burrowing, 169, 174
management, 88, 259, 274
    challenges, 275
    See also conservation; *see also under* individual species
maple (*Acer saccharum*), 113
marabou (*Leptoptilos crumenifer*), 10
marmot, yellow-bellied (*Marmota flaviventris*), 169
MaxEnt, 156
MCDA. *See* Multi-Criteria Decision Analysis
Melbourne, Australia, 59, 66, 153, 155, 159
merlin (*Falco columbarius*), 21*t*, 41*t*, 54–55, 200
    diet, 26
    nests, 24
    ranges, 21*t*, 22
merlin, Richardson's (*Falco columbarius richardsonii*), 8
mesquite (*Prosopis* spp.), 56, 80–81, 128
migration, 8, 11, 13, 95, 100, 103
    reciprocal, 11
    study methods, 158
    See also non-migratory behavior
Migratory Bird Treaty Act, 94, 208, 221, 240

Milwaukee, Wisconsin, 25, 28, 68, 95–97, 99, 263, 264
mitigation, 171, 201, 207
mitigation sites, 267
mobbing, 86–87, 134
mockingbird, northern (*Mimus polyglottus*), 134
modeling, 66–67, 156–58
molecular markers, 67
monogamy, 127, 144. *See also* fidelity: mate
Montreal, Quebec, 8, 181, 260, 261, 262
moose (*Alces americanus*), 274
mortality
    anthropogenic, 19, 52, 56, 58, 200–208, 202*f*, 206*f*, 265, 275
    exurban, 199
    of nestlings, 27, 59, 206–7
    See also injury; *see also under* individual species
mosquitoes (Culicidae spp.), 204, 205, 264
mouse, house (*Mus musculus*), 26, 45
    bait, 253
mowing, 222, 265–66
Multi-Criteria Decision Analysis, 157–58
Mycoplasma gallisepticum, 264
mythology, xi

National Audubon Society, 38
National Environmental Policy Act, 267
National Wildlife Rehabilitator's Association, 230, 231
Native Americans, xi, 240
nearest-neighbor distance, 85
NEPA. *See* National Environmental Policy Act
nest boxes, 24, 25, 140, 148, 161, 184*f*, 268
nest helpers, 97
nesting sites, 13, 18, 59, 140, 206, 265
    bridges, 261, 268
    cavities, 6, 19, 26, 139–40, 142, 148, 152, 155
    corvid, 8
    ledges, 53, 187, 260–62
    platforms, 25, 223, 268
    safety, 5–6
    selection, 23–24, 114, 128, 170, 183, 219
    trees, 80, 94, 95, 111, 114, 140, 155
nests
    defense of, 28, 110, 117, 217, 222, 250–51, 262
    density, 55, 85, 97, 99, 134
    failure, 27, 56, 84
    removal, 255, 262

reuse, 82
success, 53, 55, 83, 101
See also under individual species
New Mexico, 53, 80–81, 87, 89, 95, 99, 128–29,
134, 170, 172
New York (state), 55, 139, 147
New York City, New York, 3, 260, 261
Central Park, xii, 3
nonmigratory behavior, 8, 11, 95, 138, 205
novel features
environments, 64–65, 97
organisms, 58
processes, 52
situations, 58
structures, 24, 52, 58
NWRA. *See* National Wildlife Rehabilitator's
Association

oak (*Quercus* spp.), 80, 113, 128
oak, coast live (*Quercus agrifolia*), 114
oak, shinnery (*Quercus havardii*), 79
oak, willow (*Quercus phellos*), 140
Ohio, 25, 55, 111, 113–15, 115*f*, 118–20. *See
also* Cincinnati, Ohio
"One Health," 235
Operation High Roller, 186
Oregon, 138, 186, 189. *See also* Portland,
Oregon
osprey (*Pandion haliaetus*), 41*t*, 203, 206*f*, 214,
216, 237
nests, 25, 53–54, 206*f*, 207, 220*f*
outreach, 238, 243, 255, 262. *See also under*
conservation
owl, barn (*Tyto alba*), 45*t*, 56, 251
diet, 26, 45*t*
nests, 23, 25, 56
success, 56
owl, barred (*Strix varia*), 21*t*, 25, 45*t*, 66, 138,
279*f*
aggression, 147–48
conservation, 148
diet, 26–27, 45*t*, 142, 143*t*, 144
dispersal, 146, 147
habitat, 138–39
mortality, 146, 148
nests, 23, 140, 142
predation, 147
ranges, 21*t*, 22, 141, 142
as raptor predator, 147
and red-shouldered hawks, 146–47
reproduction, 144–45
and spotted owls, 148

success, 145
survival, 145–46
owl, burrowing (*Athene cunicularia*), 28, 29,
45*t*, 135, 166, 167*f*, 168*f*
burrows, 53, 55–56, 166–67, 167*f*, 168, 170,
171, 172*f*, 174
conservation, 171, 174–75
density, 173–74, 266
diet, 28, 45*t*, 56, 66, 170–71
dispersal, 173
habitat, 167, 169, 170, 265, 267
management, 174–75
mortality, 173
population, 167–68, 266
ranges, 174
reproduction, 172–73, 266
success, 172–73
survival, 56, 173
tolerance, 168, 171, 267
owl, great grey (*Strix nebulosa*), 67
owl, great horned (*Bubo virginianus*), 21*t*, 45*t*,
135, 282
diet, 26, 37, 45*t*
habitat, 20
nests, 25
ranges, 21*t*, 22
as raptor predator, 84, 101, 118–19, 131,
147, 182, 220–21
owl, long-eared (*Asio otus*), 20, 21*t*, 22
owl, powerful (*Ninox strenua*), 59, 65, 66, 152,
153*f*, 159*f*, 277
conservation, 161–62
detection, 155–58
diet, 154
dispersal, 157
distribution, 154
habitat, 157, 159
management, 161–62
nests, 155
population, 154
ranges, 158
reproduction, 155, 160
roosts, 155
tracking, 158
owl, short-eared (*Asio flammeus*), 276*f*, 276
owl, spotted (*Strix occidentalis*), 156
owl, spotted, Mexican (*Strix occidentalis
lucida*), 281*f*
owl, spotted, northern (*Strix occidentalis
caurina*), 148
owl, tawny (*Strix aluco*), 26, 45*t*, 156
owl pellets, 156, 280, 282

owls, 280
  nests, 25–26, 280
  See also individual species

palm, fan (*Washingtonia filifera*), 114
palo verde (*Cercidium* spp.), 128
palo verde, yellow (*Parkinsonia microphylla*),
  56
Parker, James, 80
PBDEs. *See* polybrominated diphenyl
  ethers
PCBs. *See* polychlorinated biphenyls
PCDDs. *See* polychlorinated
  dibenzo-p-dioxins
PCDFs. *See* polychlorinated dibenzofurans
pecan (*Carya illinoinensis*), 81
pedestrians, 87–90, 135, 188, 251
persecution, 8, 84, 88, 94, 101, 131, 186–87,
  207, 261
persistence, 52
pesticides, 8, 56, 58, 180–81, 207. *See also*
  poison; rodenticides; *individual*
  pesticides
pests, 26, 216
pets, 219, 222, 251, 259
pigeon, rock (*Columba livia*), 6, 12, 21, 26,
  186, 261
pine (*Pinus* spp.), 102, 114, 128
pine, Aleppo (*Pinus halepensis*), 56, 134
pine, white (*Pinus strobus*), 95
pine plantations, 102, 114
placement
  captive, 234
  postrehabilitation, 235
plasticity. *See* adaptability
play, 147–48
poison, 119, 121, 131, 188, 261, 264, 268.
  See also contaminants; pesticides;
  rodenticides
polyandry, 127
polybrominated diphenyl ethers, 203, 264
polychlorinated biphenyls, 181, 203
polychlorinated dibenzofurans, 203
polychlorinated dibenzo-p-dioxins, 203
polygyny, 28, 97–98, 127, 144
population, 53, 68
  density, 57, 205
  growth, 58
  See also under individual species
Portland, Oregon, 186, 187
possum, common brushtail (*Trichosurus*
  vulpecula), 154

possum, common ringtail (*Pseudocheirus*
  peregrinus), 154
poultry, 205, 219, 222, 251
pox, avian (Poxviridae spp.), 205
prairie dog, black-tailed (*Cynomys*
  ludovicianus), 169
predation, 8, 55, 83–84, 86, 99, 101, 206, 276.
  See also under individual species
presentations, live. *See* education: live
  presentations
prey
  amphibian, 81, 116, 142, 144
  avian, 26–27, 45–46, 57, 81, 93, 116, 129,
    135, 142, 144, 170, 182, 249
  fish, 116
  insect, 8–9, 12, 28, 37, 81, 86, 170–71
  invertebrate, 44, 116, 142
  mammalian, 27, 28, 37, 44, 81, 93, 116, 129,
    135, 142, 144, 170, 222
  management of, 222
  marsupial, 154
  reptile, 37, 81, 116, 129, 142, 144
  synanthropic, 26
  See also diet; food
productivity, 55, 59. *See also under individual*
  species
public awareness, 89, 229, 241, 243, 262, 269,
  280, 284
public relations, 268

rabbit, cottontail (*Sylvilagus* spp.), 129
rabbit, European (*Oryctolagus cuniculus*), 130,
  263
raccoon (*Procyon lotor*), 84, 101, 119, 182
radio tagging, 142
ranges. *See under individual species*
Raptor Center, 231, 236, 240, 242
Raptor Lab, 242
Raptor Med, 236
raptor medicine, 236
Raptor Research Foundation, xii
rat (*Rattus* spp.), 18, 45
rat, brown (*Rattus norvegicus*), 26, 144
refuse. *See* dumps; garbage; landfills
rehabilitation, 243, 254, 269
  data gathering, 235–36
  legal requirements, 230–31
  of migratory birds, 231
reintroduction, 51
release
  postrehabilitation, 232, 234, 235t
  soft (*see* hacking)

relocation, passive, 266–68
reproduction, 68, 83
  success, 19, 27, 145
  See also monogamy; polyandry; polygyny;
    see also under individual species
reproductive rate, 19, 53, 65, 99–101, 118
research
  clinical, 236
  and education, 216
  genetic, 67–68
  methodology, 20, 65, 66–68, 69, 154, 156–61
  molecular, 67
  and rehabilitation, 235–36
  subjects of, 280, 282
roadrunner, greater (*Geococcyx californianus*),
    275
robin, American (*Turdus migratorius*), 93
rodenticides, 110, 203, 204, 222, 232, 268
  anticoagulant, 56, 121, 204
  second-generation anticoagulant, 148, 204,
    236
  See also pesticides
roosting sites, 155

saguaro (*Carnegiea gigantean*), 128
saltcedar (*Tamarix* spp.), 80
San Antonio, Texas, 250, 252, 254, 255
Santa Clara Valley Audubon Society, 267
satellites, 67, 158, 237
scavenging, 10, 27, 116, 216
screech-owl (*Megascops* spp.), 147
screech-owl, eastern (*Megascops asio*), 19, 54,
    156
  nests, 24, 25
SDM. *See* Species Distribution Modeling
Seattle, Washington, 139, 147, 187
sexual dimorphism, 93–94, 152
SGARs. *See* rodenticides: second-generation
    anticoagulant
shelterbelts, 80, 85
shooting, 84, 101, 207, 232
slaughterhouses. *See* abattoirs
Snake River Birds of Prey Conservation Area,
    4
soft release. *See* hacking
Spain, 3, 8, 11, 12, 24, 27, 35
sparrow, house (*Passer domesticus*), 8, 26, 93,
    95, 99, 263
sparrowhawk, Eurasian (*Accipiter nisus*), 6,
    26–27
specialists, 12, 152, 200
Species Distribution Modeling, 67

sprawl, 9, 52, 120, 159, 161–62, 283
squirrel (Sciuridae spp.), 142
squirrel, eastern gray (*Sciurus carolinensis*),
    119
squirrel, ground (*Otospermophilus beecheyi*),
    265
squirrel, ground (Sciuridae spp.), 129
starling, European (*Sturnus vulgaris*), 6, 99,
    188, 263
starvation, 27, 59, 98, 232
Stevens Point, Wisconsin, 96, 99, 100*f*, 264,
    278
substrates, 24–26, 170, 260
  artificial, 54, 171, 202, 223
  natural, 53–54
Sun Life Building, 8, 181, 260–62
survival, 19, 53, 55, 57. *See also under*
    individual species
swift, Vaux's (*Chaetura vauxi*), 188
sycamore (*Platanus occidentalis*), 113, 140
sycamore, western (*Platanus racemosa*), 114
symbiosis, 4–6. *See also* commensalism
symbolism, xi–xii

telemetry, 237
temperature, 95
tern, least, California (*Sternula antillarum*
    browni), 220
territoriality, 13, 127, 147, 156, 262
Texas, 39*t*, 54, 55, 80–83, 85, 88, 128, 134, 138,
    169–74, 181, 186, 237, 246–56. *See also*
    Lubbock, Texas; San Antonio, Texas
Texas A&M AgriLife Extension Service, 255
Texas Master Naturalist Program, 255–56
Texas Parks and Wildlife Department, 247–56
Texas Urban Wildlife Program, 255
time-activity budgets, 29
tolerance, 64–66, 86. *See also under individual*
    species
towers, communication, 219, 220*f*
toxins. *See* contaminants; pesticides; poison
tracking, 66–67
  GPS, 158–59, 159*f*, 161
  satellite, 158
  VHF, 158
translocation, 88–89, 90, 117, 224
transmitters, 158
trapping, 131, 224, 230, 232, 236
trash. *See* dumps; garbage; landfills
TRC. *See* Raptor Center
Trichomonas gallinae, 27, 59, 102, 188, 205
trichomoniasis, 59, 102, 131, 205, 236, 264

Tucson, Arizona, 20, 21*t*, 27, 36, 47, 56–57,
　　59, 95, 97–99, 102, 127–32, 134–35, 172*f*,
　　207, 264, 278
turkey, wild (*Meleagris gallopavo*), 274, 284

United States Geological Survey, 84, 237
urban-adaptors, 64
urbanization
　　benefits, 259, 275
　　of burrowing owls, 167–70
　　and evolution, 13
　　of landscape, 26, 36–37
　　of peregrine falcons, 8, 181, 260
　　as process, 273
　　of raptors, 5, 12, 19
　　regional, 56–57, 60
　　and resources, 54
　　risks, 275
　　of wildlife, 274
Urban Wildlife Program, 247
USGS. *See* United States Geological Survey

vagrancy, 120
vegetation, 27, 55, 57, 79–80, 130, 140, 144
　　native, 53, 56, 128, 135
　　nonnative, 56, 114, 116, 135
Very High Frequency radio, 158
Very High Frequency telemetry, 237
veterinarians, 236
VHF. *See* Very High Frequency radio; Very
　　High Frequency telemetry
Victoria, British Columbia, 95–96, 99, 158,
　　263–64
viruses, 205
vocalizations, 97, 134, 147, 156
voles (*Microtus* spp.), 28, 44
volunteers, 231–32, 255–56, 261
vulture, black (*Coragyps atratus*), 11–12, 40,
　　41*t*, 250, 252
　　nests, 253*f*

vulture, Egyptian (*Neophron percnopterus*), 9,
　　10–11, 12–13
vulture, griffon (*Gyps fulvus*), 11, 12–13
vulture, hooded (*Necrosyrtes monachus*), 9–11,
　　12
vulture, turkey (*Cathartes aura*), 11–13, 40, 41*t*
vulture, white-rumped (*Gyps bengalensis*), 9
vultures, 3, 29
　　human conflict, 255
　　symbiosis, 4–5, 9
　　See also individual species

warning signs, 89, 90*f*, 117*f*, 222
Washington (state), 138–39, 169, 173, 184,
　　186, 189. *See also* Seattle, Washington
Washington, DC, xii, 139
"watchable wildlife," xii, 166
wattle (*Acacia* spp.), 155
waxwing, Bohemian (*Bombycilla garrulus*), 8
webcams, 216, 262, 269
West Nile virus (Flaviviridae spp.), 121, 206,
　　264
wildlife agencies, 247
Wildlife Rehabilitation MD, 236
WILD-ONe, 236
willow (*Salix* spp.), 114
willow, Goodding's black (*Salix gooddingii*),
　　114
windbreaks, 80–81, 114
wind farms, 201
Wisconsin, 55, 57, 94–95, 98–99, 100*f*,
　　101–3, 145, 260, 263–64, 278. *See also*
　　Milwaukee, Wisconsin; Stevens Point,
　　Wisconsin
WNV. *See* West Nile virus
wolf (*Canis lupus*), 275
woodrat (*Neotoma* spp.), 129

zoonotic diseases. *See* disease: zoonotic
zoos, 239